T0190369

Advances in Intelligent Systems and Computing

Volume 666

Series editor

Janusz Kacprzyk, Polish Academy of Sciences, Warsaw, Poland
e-mail: kacprzyk@ibspan.waw.pl

The series "Advances in Intelligent Systems and Computing" contains publications on theory, applications, and design methods of Intelligent Systems and Intelligent Computing. Virtually all disciplines such as engineering, natural sciences, computer and information science, ICT, economics, business, e-commerce, environment, healthcare, life science are covered. The list of topics spans all the areas of modern intelligent systems and computing.

The publications within "Advances in Intelligent Systems and Computing" are primarily textbooks and proceedings of important conferences, symposia and congresses. They cover significant recent developments in the field, both of a foundational and applicable character. An important characteristic feature of the series is the short publication time and world-wide distribution. This permits a rapid and broad dissemination of research results.

More information about this series at http://www.springer.com/series/11156

Rituparna Chaki · Agostino Cortesi
Khalid Saeed · Nabendu Chaki
Editors

Advanced Computing and Systems for Security

Volume Five

 Springer

Editors
Rituparna Chaki
A. K. Choudhury School
  of Information Technology
University of Calcutta
Kolkata, West Bengal
India

Agostino Cortesi
Dipartimento di Scienze Ambientali,
  Informatica e Statistica
Università Ca' Foscari
Venice
Italy

Khalid Saeed
Faculty of Computer Science
Bialystok University of Technology
Bialystok
Poland

Nabendu Chaki
Computer Science and Engineering
University of Calcutta
Kolkata, West Bengal
India

ISSN 2194-5357   ISSN 2194-5365   (electronic)
Advances in Intelligent Systems and Computing
ISBN 978-981-10-8179-8   ISBN 978-981-10-8180-4   (eBook)
https://doi.org/10.1007/978-981-10-8180-4

Library of Congress Control Number: 2017964593

Printed on acid-free paper

This Springer imprint is published by the registered company Springer Nature Singapore Pte Ltd. part of Springer Nature
The registered company address is: 152 Beach Road, #21-01/04 Gateway East, Singapore 189721, Singapore

Preface

The Fourth International Doctoral Symposium on Applied Computation and Security Systems (ACSS 2017) took place on March 17–19, 2017, in Patna, India.

University of Calcutta, along with Ca' Foscari University of Venice, Bialystok University of Technology, and Warsaw University of Technology collaborated to make ACSS 2017 a grand success. Around 40 participants from a multitude of institutions have taken part in a highly interactive discussion for 3 days, resulting in a cumulative experience of research idea exchange.

The post-conference book series are indexed by ISI Web of Sciences. The sincere effort of the program committee members coupled with ISI indexing has drawn a large number of high-quality submissions from scholars all over India and abroad. The Technical Program Committee for the symposium selected only 21 papers for publication out of 70 submissions.

The papers mainly covered the domains of computer vision and signal processing, biometrics-based authentication, security for Internet of Things, analysis and verification techniques, security in mobile and cloud scenarios, large-scale networking, remote health care, distributed systems, software engineering, cloud computing, privacy and confidentiality, access control, big data and data mining, android security.

The technical program was organized into six topical sessions each day. The sessions started with a keynote lecture on a pertinent research issue by an eminent researcher/scientist. This was followed by short, to-the-point presentations of the technical contributions. At the end of each session, the session chair handed over the suggestions for improvement pertaining to each paper. The sessions also saw lively discussions among the members of the audience and the presenters.

The post-conference book includes the presented papers in enhanced forms, based on the suggestions of the session chairs and the discussions following the presentations. Each of the accepted papers had undergone a double-blind review process. During the presentation, every presented paper was evaluated by the concerned session chair, an expert in the related domain. As a result of this process, most of the papers were thoroughly revised and improved, so much so that we feel that this book has become much more than a mere post-workshop proceedings volume.

We would like to take this opportunity to thank all the members of the Technical Program Committee and the external reviewers for their excellent and time-bound review works. We thank all the sponsors who have come forward toward the organization of this symposium. These include Tata Consultancy Services (TCS), Springer India, ACM India. We appreciate the initiative and support from Mr. Aninda Bose and his colleagues in Springer for their strong support toward publishing this post-symposium book in the series "Advances in Intelligent Systems and Computing". Last but not least we thank all the authors without whom the symposium would not have reached up to this standard.

On behalf of the editorial team of ACSS 2017, we sincerely hope that this book will be beneficial to all its readers and motivate them toward further research.

Kolkata, India   Rituparna Chaki
Venice, Italy   Agostino Cortesi
Bialystok, Poland   Khalid Saeed
Kolkata, India   Nabendu Chaki

Contents

Part III   Pattern Recognition

About the Editors

Rituparna Chaki is Professor of Information Technology at the University of Calcutta, India. She received her Ph.D. from Jadavpur University in India in 2003. Before this she completed her B.Tech. and M.Tech. in Computer Science and Engineering from University of Calcutta in 1995 and 1997, respectively. She has served as a System Executive in the Ministry of Steel, Government of India, for 9 years, before joining the academics in 2005 as a Reader of Computer Science and Engineering in the West Bengal University of Technology, India. She is with the University of Calcutta since 2013. Her areas of research include optical networks, sensor networks, mobile ad hoc networks, Internet of Things, data mining. She has nearly 100 publications to her credit. She has also served in the program committees of different international conferences. She has been a regular Visiting Professor in the AGH University of Science and Technology, Poland, for last few years. She has co-authored a couple of books published by CRC Press, USA.

Agostino Cortesi Ph.D. is a Full Professor of Computer Science at Ca' Foscari University, Venice, Italy. He served as Dean of the Computer Science Studies, as Department Chair, and as Vice-Rector for quality assessment and institutional affairs. His main research interests concern programming languages theory, software engineering, and static analysis techniques, with particular emphasis on security applications. He published more than 110 papers in high-level international journals and proceedings of international conferences. His h-index is 16 according to Scopus and 24 according to Google Scholar. He served several times as member (or chair) of program committees of international conferences (e.g., SAS, VMCAI, CSF, CISIM, ACM SAC), and he is in the editorial boards of the journals "Computer Languages, Systems and Structures" and "Journal of Universal Computer Science." Currently, he holds the chairs of "Software Engineering," "Program Analysis and Verification," "Computer Networks and Information Systems," and "Data Programming."

Khalid Saeed is a Full Professor in the Faculty of Computer Science, Bialystok University of Technology, Bialystok, Poland. He received B.Sc. in Electrical and Electronics Engineering in 1976 from Baghdad University in 1976 and M.Sc. and Ph.D. from Wroclaw University of Technology, Poland, in 1978 and 1981, respectively. He received his D.Sc. (Habilitation) in Computer Science from Polish Academy of Sciences in Warsaw in 2007. He was a Visiting Professor of Computer Science at Bialystok University of Technology, where he is now working as a full professor. He was with AGH University of Science and Technology in the years 2008–2014. He is also working as a Professor in the Faculty of Mathematics and Information Sciences at Warsaw University of Technology. His areas of interest are biometrics, image analysis and processing, and computer information systems. He has published more than 220 research papers, and edited 28 books, journals and conference proceedings, and authored 10 text and reference books. He supervised more than 130 M.Sc. and 16 Ph.D. theses. He gave more than 40 invited lectures and keynotes at different conferences and universities in Europe, China, India, South Korea, and Japan on biometrics, image analysis and processing. He received more than 20 academic awards. He is a member of more than 20 editorial boards of international journals and conferences. He is an IEEE Senior Member and has been selected as IEEE Distinguished Speaker for 2011–2016. He is the Editor-in-Chief of International Journal of Biometrics with Inderscience Publishers.

Nabendu Chaki is a Professor in the Department Computer Science and Engineering, University of Calcutta, Kolkata, India. He did his first graduation in Physics from the legendary Presidency College in Kolkata and then in Computer Science and Engineering from the University of Calcutta. He has completed his Ph. D. in 2000 from Jadavpur University, India. He is sharing six international patents including four US patents with his students. He has been quite active in developing international standards for Software Engineering and Cloud Computing as a member of Global Directory (GD) for ISO/IEC. Besides editing more than 25 book volumes, he has authored 6 text and research books and has more than 150 Scopus Indexed research papers in journals and international conferences. His areas of research interests include distributed systems, image processing, and software engineering. He has served as a Research Faculty in the Ph.D. program in Software Engineering in US Naval Postgraduate School, Monterey, CA. He is a visiting faculty member for many universities in India and abroad. Besides being in the editorial board for several international journals, he has also served in the committees of over 50 international conferences. He is the founder Chair of ACM Professional Chapter in Kolkata.

Part I
Biometrics

Hand Biometric Verification with Hand Image-Based CAPTCHA

Asish Bera, Debotosh Bhattacharjee and Mita Nasipuri

Abstract An approach for hand biometric recognition with the hand image-based CAPTCHA verification is presented in this paper. A new method for CAPTCHA generation is implemented based on the genuine and fake hand images which are embedded in a complex textured color background image. The $HANDCAPTCHA$ is a useful application to differentiate between the human and automated scripts. The first level of security is achieved by the $HANDCAPTCHA$ against the malicious threats and attacks. After solving the $HANDCAPTCHA$ correctly, the identity of a person is authenticated based on the contact-less hand geometric verification approach in the second level. A set of 300 unique $HANDCAPTCHA$ is created randomly and solved by at least 100 persons with the accuracy of 98.34%. Next, the left-hand images of the legitimate users are normalized, and sixteen geometric features are computed from every normalized hand. Experiments are conducted on the 200 subjects of the Bosporus left-hand database. Classification accuracy of 99.5% has been achieved using the kNN classifier, and the equal error rate is 3.93%.

Keywords CAPTCHA · $HANDCAPTCHA$ · Biometrics · Verification
Equal error rate

A. Bera (✉)
Department of Computer Science and Engineering,
Haldia Institute of Technology, Haldia 721657, India
e-mail: asish.bera@gmail.com

D. Bhattacharjee · M. Nasipuri
Department of Computer Science and Engineering, Jadavpur University,
Kolkata 700032, India
e-mail: debootsh@ieee.org

M. Nasipuri
e-mail: mitanasipuri@gmail.com

© Springer Nature Singapore Pte Ltd. 2018
R. Chaki et al. (eds.), *Advanced Computing and Systems for Security*,
Advances in Intelligent Systems and Computing 666,
https://doi.org/10.1007/978-981-10-8180-4_1

1 Introduction

Biometrics is a convenient technology for secure human authentication in various commercial, government, and forensic applications [1]. Biometrics is considered as the reliable alternative to the password-based authentication from unauthorized access through the Internet [2]. Furthermore, various types of multimedia contents such as a photograph, audio, and video in social networking Web sites are increasing which cause illegitimate access to the system [3]. The human hand is one such reliable biometric trait that has been globally accepted for automated human recognition to minimize security threats [4, 5]. Nowadays, handheld devices like a smartphone with various potential biometric verification facilities are globally popular. In recent years, Web-based applications are unavoidable in different circumstances. These systems need to verify whether the user who wants to access a service is a person or an intelligent computer system. Automated malicious program known as the 'bots' creates a fake identity of a human automatically for the identity theft [6]. Biometrics is an efficient technique that essentially provides better security against machine-based attacks.

Hand geometry is a viable alternative when less data storage is a requirement for identity verification such as automated attendance system, or boarder control of different nations [7]. Different attributes of hand and fingers are computed for individualization of a legitimate person. Hand biometric systems are mainly accepted for verification task. Geometric measurements [8] and shape-based [4, 9] features are investigated in hand biometrics. Shape-based hand recognition approaches based on the modified Hausdorff distance (MHD) and independent component analysis (ICA) are described in [9]. The ICA tool has been applied on normalized binary hand images to extract and summarize prototypical shape information. In another verification approach, geometric features of four fingers from 100 subjects have been used for experimentation [10]. Selection of useful features enhances the accuracy. A feature selection scheme using the Genetic Algorithm (GA) and mutual information is described in [11]. GA with a fitness function is applied for selecting discriminative hand features such as geometric and shape-based descriptors. Mutual information is applied to find out the correlation between a pair of features and to eliminate redundant features. A hand biometric-based verification method using the scale-invariant feature transform (SIFT) is presented in [12]. The SIFT features are invariant to scale changes, rotation, noise, and distortion. Research interests are also concentrating on the 3-D hand geometry, a fusion of 2-D and 3-D hand features [13], and thermal hand images [14] in the contact-free environment.

Completely Automated Public Turing Test to Tell Computers and Humans Apart, or in the acronym, CAPTCHA is designed to distinguish between legitimate users and automated scripts [15, 16]. The CAPTCHA is mainly used for Web-based, financial-based services to provide security measures against unauthorized access. The main objective of a CAPTCHA is to provide a testing method by which a human can be discriminated from an automated program. The more difficulty and randomness added to a CAPTCHA-based verification task, the more reliability can be expected to perform better against the bot attacks. The difficulty is added mainly for the test based on visualization, i.e., a normal human perception can identify the necessary objects and solve in a simpler manner within a fewer seconds. The pattern generated by a CAPTCHA should be sufficiently random so that an automated intelligent system cannot solve very easily. The patterns are created for a CAPTCHA tool mainly based on the following:

(i) Textual and numerical data which is primarily used in various Web-based applications. The reCAPTCHA is a popular tool based on text-CAPTCHA [17].

(ii) Different types of visual images such as nature, animal, human face, cartoons, and other graphical objects.

(iii) Audio- and video-based CAPTCHA testing is an alternative approach. However, the necessary memory size and bandwidth are the primary limitations.

(iv) Simple visual CAPTCHA puzzle solving is another alternative solution.

Text-based CAPTCHA is very popular in various common applications and services. Text-CAPTCHA is generated in the form of an image which is processed with several transformation and distortion so that a human can identify as a randomly generated sequence of the case-sensitive alphabets, numerical digits, and special symbols. More than hundred millions of text-CAPTCHAs are solved worldwide per day [18]. However, due to language dependency in text-CAPTCHA is not useful for every circumstance. On the other side, image-based CAPTCHA is a valuable tool for testing a human. The image-based methods based on the human faces are implemented in [15, 16]. It is a robust solution against malicious attacks. Several transformations and distortion methods are followed to generate the face-CAPTCHA and face-DCAPTCHA. However, biometric characteristics of the face have not been investigated in these works. According to our study, no work exists in the literature for biometric authentication scheme that incorporates the benefits of any type of CAPTCHA.

In this work, a new $HANDCAPTCHA$ verification method has been implemented that is based on the hand images. It enables a human to solve a CAPTCHA based on several images of hand which is generated automatically using a randomized combination of two real and other fake hand images. It also identifies the user is a human or not. Unlike only two different distortions are applied consecutively in [16], some transformation and distortion methods are successively applied to design the $HANDCAPTCHA$. After successfully solving the $HANDCAPTCHA$, the person is authenticated based on hand biometric verification approach in an unconstrained environment. The system offers two levels of stringent security for correct verification of a person by avoiding unauthorized access mechanism through the intelligent computer programs. The main contributions of this work are as follows:

(a) A hand biometric system with the additional advantages of hand image-based CAPTCHA.

(b) Various transformation and distortion methods are successively applied to enhance the robustness of the $HANDCAPTCHA$. Additionally, new transformation method is also implemented.

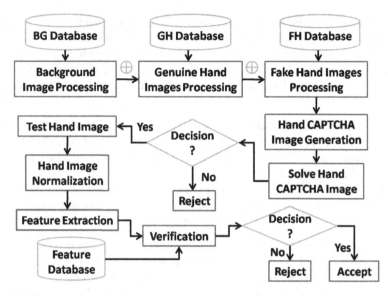

Fig. 1 Hand biometric verification methodology using *HANDCAPTCHA*

(c) Hand biometric verification is performed based on the normalized pose-invariant geometric features of the hand.

The rest of this paper is organized as follows: Sect. 2 describes the methods to generate the *HANDCAPTCHA*. Section 3 presents the steps of image preprocessing for hand normalization and feature definition. In Sect. 4, experimental results of *HANDCAPTCHA* verification and hand biometric authentication are described. The conclusion is drawn in Sect. 5.

2 *HANDCAPTCHA* Processing and Verification Scheme

The proposed framework for hand biometric verification is given in Fig. 1. It consists of mainly two parts. Firstly, the *HANDCAPTCHA* is generated and verified by a person. Secondly, the identity verification of a legitimate user is performed with the hand biometric verification method.

(i) *HANDCAPTCHA* preprocessing

Initially, the datasets are categorized into three types, namely background database (BG database, D_{BG}), genuine hand database (GH database, D_{GH}), and fake hand database (FH database, D_{FH}). These three datasets are combined to generate a *HANDCAPTCHA* image. After preprocessing the background image (I_{BG}), successive steps are followed for processing the genuine hand images (I_{GH}) and overlaid on I_{BG} image. The formation of superimposed image by overlaying a hand image

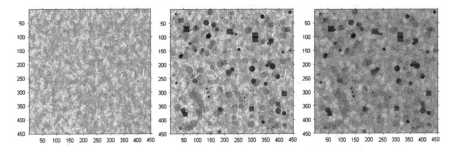

Fig. 2 Background image processing for $HANDCAPTCHA$ creation

over the I_{BG} image is denoted by $\oplus$ symbol in Fig. 1. Similarly, the fake hand images (I_{FH}) are processed and placed over the processed I_{BG} and I_{GH} images.

(a) Background image preprocessing

The dimension of $HANDCAPTCHA$ output image (I_{HC}), in terms of width (x_size) and height (y_size), is selected as 460×460 pixels. However, it can be set to any dimension in which device this application is used for verification. Similarly, image I_{BG} is fixed to the same size of I_{HC}, i.e., 460×460 pixels. The total number of I_{BG} images is denoted by N_{BG}. Firstly, one I_{BG} image is selected from the dataset of background images, D_{BG}. All the I_{BG} images are assigned to a unique label. A random number $\mathbb{N}$ is chosen which lies within N_{BG} and $\mathbb{N} \in \{1, N_{BG}\}$. Now, $\mathbb{N}$ is considered as the label index, based on which the Nth image I_{BG} is selected. Similarly, the I_{GH} and I_{FH} images are selected according to the randomly generated label index.

Now, I_{BG} undergoes few different operations with randomized values of various parameters. These are arbitrary pixel positions, size, shape, opacity, and fill color. Then, a set of 300 different pixel positions $p(x\_pos, y\_pos)$ is selected. Different object shapes mainly circle, star, and rectangle with the various sizes up to a maximum of 30 pixels are chosen automatically. RGB colors with intensities $\{0\text{--}255, 0\text{--}255, 0\text{--}255\}$ and opacity $\{0, 1\}$ are selected randomly. Now, these shapes L_{Circle}, L_{Rect} are placed on those random pixel positions. In this case, overlapping of two different shapes is allowed to avoid unnecessary complexity. Also, the morphological opening is also performed with a maximum disk size of 10, and it is also varied randomly. Finally, the salt and pepper noise with density 0.05 is added in the processed image I'_{BG}. In Algorithm 1, steps 1–8 are followed for processing the background image, and it results in I'_{BG} (Fig. 2).

Algorithm 1: *HANDCAPTCHA* Generation

Input: *total number of images (N_T), background image (I_{BG}), number of genuine hand images (N_{GH}), number of fake hand images (N_{FH}).*

Output: *HANDCAPTCHA image: I_{HC}*

1.   **Initialize:** $N_T = N_{GH} + N_{FH} = 2 + 7 = 9$; *x_size = y_size = 460 pixels.*
2. $I_{BG} \leftarrow Random(label\_index, 1, x\_size, y\_size, D_{BG})$ //* **background image processing** *//
3. $L_{Circle} \leftarrow Random(150, x\_pos, y\_pos)$
4. $L_{Rect} \leftarrow Random(150, x\_pos, y\_pos)$
5. $I'_{BG} \leftarrow I_{BG} + Random(L_{Circle}, size, RGB\ color, opacity)$
6. $I'_{BG} \leftarrow I'_{BG} + Random(L_{Rect}, size, RGB\ color, opacity)$
7. $I'_{BG} \leftarrow MorphologicalOpen(I'_{BG}, disk, disk\_radius)$
8. $I'_{BG} \leftarrow AddNoise(I'_{BG}, "salt\ and\ peeper", noise\_density)$
9. $I_{HC} \leftarrow I'_{BG}$
10.   For *i=1:2*   //* **genuine hand image processing** *//
11. $I_{GH} \leftarrow Random(class\ label\_index, i, GH_{x\_size}, GH_{y\_size}, D_{GH})$
12. $L_G \leftarrow Random(x\_pos, y\_pos, length, width)$
13. $\theta(i) \leftarrow Random(Rotation, \pm angle)$
14. $I'_{GH} \leftarrow Rotate(I_{GH}, \theta(i))$
15. $I_{HC} \leftarrow I_{HC} \oplus PositionMapping(I'_{GH}, L_G)$
16.   End for
17.   For *i=1:7*   //* **fake hand image processing** *//
18. $I_{FH} \leftarrow Random(label\_index, i, FH_{x\_size}, FH_{y\_size}, D_{FH})$
19. $L_F \leftarrow Random(x\_pos, y\_pos, length, width)$
20. $\theta(i) \leftarrow Random(Rotation, \pm angle)$
21. $I'_{GH} \leftarrow Rotate(I_{FH}, \theta(i))$
22. $I_{HC} \leftarrow I_{HC} \oplus PositionMapping(I'_{GH}, L_F)$
23.   End for
24. $I_{HC} \leftarrow AlphaBlend(I_{HC}, I'_{BG}, opacity)$
25.   Return I_{HC}
26.   End

(b) Genuine and fake hand image preprocessing

After processing I'_{BG} image, two hand images (I_{GH}) of a genuine person and maximum seven different fake hand images (I_{FH}) are chosen and placed over I'_{BG} image at random spatial locations. Image I'_{BG} is divided into nine blocks, i.e., 3×3 blocks, and a label is assigned to each block where the I_{GH} and I_{FH} hand images are mapped and placed. The label is assigned vertically, according to the column-major order in the I'_{BG}. Now, any image of either I_{GH} or I_{FH} is put on any block. In the case of every I_{GH}, the following steps are followed.

The class label of I_{GH} is chosen from the database D_{GH} along with the dimension, i.e., $GH_{x\_size} \times GH_{y\_size}$. The size, i.e., *length $\times$ width* of every image is also variable, and it should not exceed 150×150 pixels. Now, I_{GH} is rotated with an arbitrary angle $\pm\theta$ in either clockwise ($-\theta$) or counterclockwise (θ) direction. In the case of I_{GH}, the real hand area is segmented from the darker background based on intensity variation at the hand boundary region. For this purpose, the region of interest (ROI) of the hand is selected, and the background is neglected. After selecting the ROI, image I_{GH} is mapped to the respective positional block and placed over the I'_{BG} according to its length and width. Mapping is required to maintain the spaces between any two hand images. Moreover, mapping also maintains a least amount of area around

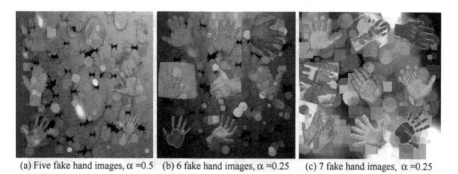

(a) Five fake hand images, α =0.5   (b) 6 fake hand images, α =0.25   (c) 7 fake hand images, α =0.25

Fig. 3 Sample outcomes of the $HANDCAPTCHA$ with different $N_{FH} = 5$, 6, and 7

the boundary that should not be covered by any hand image. For genuine image processing steps, 10–16 of Algorithm 1 are followed. Similar operations are followed by the I_{FH} images from steps 17–23.

At last, the alpha-blending is applied on the resultant images I_{HC} and I'_{BG} with a variable amount of opacity (α) to make a combined $HANDCAPTCHA$ image. The time complexity of the algorithm depends on the dimension of I_{HC}, and it can be represented as $\mathbf{O}(x\_size \times y\_size)$. Sample outputs of the $HANDCAPTCHA$ are shown in Fig. 3. It can be observed that the value of α is higher in Fig. 3a, whereas in the two other cases it is kept lower because higher value of α creates visualization error in some cases. Finally, a challenging I_{HC} image is ready for collecting user response.

(c) *HANDCAPTCHA* verification method

The user is asked to solve the $HANDCAPTCHA$ image, I_{HC}. The user has to find out two hand images which are collected from the same person and are genuine, i.e., I_{GH}. This can be solved by observing I_{HC} image and finding out the two similar I_{GH} images by mentioning their positions in a text field provided in a GUI-based application containing I_{HC}. Here, the text field is used to collect the user responses. If a user can identify the two I_{GH} images successfully, then he/she is allowed for hand biometric verification method, otherwise not. Every unsuccessful attempt implies the user is rejected. The experimental details are described in Sect. 4.

3 Hand Image Normalization and Feature Definition

Hand image normalization is the most important stage for defining the hand template in a contact-less image acquisition environment. The preprocessing of a color hand image follows successive sub-steps specifically, the darker background elimination, rotation of the main hand contour component, removal of wrist irregularities, and finger tip–valley localization. The major outcomes of the subsequent steps are depicted in Fig. 4.

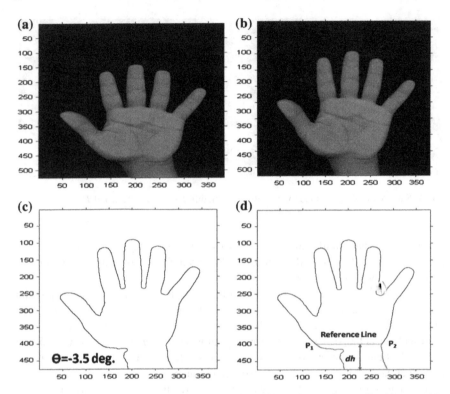

Fig. 4 **a** Color image, **b** grayscale image after rotation, **c** binary image, **d** normalized image with wrist removal

(a) Basic preprocessing of hand image

Firstly, a color hand image H_C is converted into the grayscale image H_G. Then, segmentation is applied to H_G according to optimal gray-level thresholding by the *Otsu* method [19], and a median filter is applied to eliminate the noise effect. Next, H_G is converted into a binary image, and hand contour is determined by the *Canny's* edge detector, H_B. The hand contour image includes either the main contour component as a single connected component or the main contour component along with few minor components due to hand accessories like wristwatch. Thus, the largest binary hand component H_L is selected, and all smaller components are eliminated. Now, the hand region is cropped to extract real hand area, denoted by H_R. The hand region H_R is determined by considering the minimum bounding rectangle around its boundary. Now, rotation with an angle $\pm\theta$ degree is applied on H_R so that it becomes perpendicular with the Cartesian x-axis. For this purpose, an ellipse is fitted over the hand contour so that its major axis passes through the centroid of the hand contour. The centroid is defined as

$$(x_c, y_c) = (m_{1,0}/m_{0,0}, m_{0,1}/m_{0,0}) \tag{1}$$

The angle $(\pm\theta)$ between the coordinate y-axis and the major axis of the fitted ellipse is considered for rotating the H_R image. Consequently, rotation causes the distortion and intensity variation in hand shape which has been avoided by the bilinear interpolation. Morphological operators are applied to smooth the contour image. The resultant image is denoted by H_M. The angle of rotation is given as

$$\theta = 0.5\, tan^{-1}\left(\frac{2m_{1,1}}{m_{2,0} - m_{0,2}}\right), \tag{2}$$

where $m_{i,j}$ is the moment of the image.

In the wrist region of the main hand component, some geometrical irregularities exist which change from person to person and degrade the accuracy of feature computation. Smoothing is a technique which eliminates this type of irregularity. For this purpose, a reference line at a distance dh is considered whose lower portion contains the lower palm and wrist region. The value of dh is determined approximately as the 1/5th of the total length of the main component. The leftmost (P_1) and rightmost (P_2) nonzero pixels are the two endpoints of the reference line P_1P_2, shown in Fig. 4d.

(b) Fingertips and valleys localization

Localization of finger tip and valley points is an important task for geometric feature computation. The radial distance method is used for finding extreme points of fingers on the middle point (P_{Mid}) of the reference line P_1P_2. At first, identify the point that is the maximum distance apart from P_{Mid}. Based on the normal human hand anatomic structure, this point is considered as the tip of the middle finger according to the radial distance. Now, based on the tip of the middle finger, the extreme points of the other fingers are located. In the case of the left hand, the left side of middle finger contains the tips of thumb and index fingers along with their associated valleys. The right side of middle finger consists of the tips and valleys of ring and little fingers of the left hand. For the right-hand tips and valleys, the reverse scenario takes place. The tips and valleys of other fingers are located based on the radial distance map, particularly the maximum Euclidean distances from P_{Mid} and the angles formed with the reference line. A detailed description of locating the extreme points using the radial distance basis is presented in [9]. Finally, the features are computed with these finger tip–valley points.

(c) Feature extraction

Geometric features are easier to compute and require less memory to store the feature template in the database. Most of the features are computed based on the Euclidean distances from the centroid of the normalized hand. It should be noted that tip of the thumb is avoided for feature computation. As the thumb is the most flexible finger, its distance varies mainly due to inter-finger spacing conditions. The features are pictorially shown in Fig. 5.

Now, the feature set F includes 16 different features $F = \{f_i\}_{i=1}^{16}$, and the magnitudes of the features are in various scale. It causes a feature f_i with higher value to dominate over another feature f_j. For example, the area of a hand is in the range

Fig. 5 Feature
representation with labels

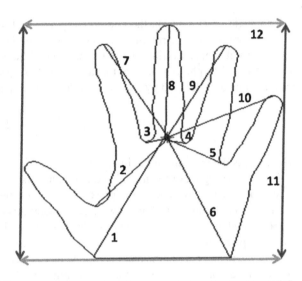

Feature definition

Altogether 16 different features are defined as
- The distances from the centroid (x_c, y_c) to ten different extreme points on hand contour are calculated as $D_i = sqrt[\,(x_c - x_i)^2 + (y_c - y_i)^2]$, where (x_i, y_i) is the ith extreme point on the hand contour.

- Length and width of normalized hand shape.

- The area of the normalized hand, major-axis length, and minor-axis length of the ellipse fitted over the hand, equivalent diameter of a circle with the same area as the hand, computed as $(4 \times AREA/\pi)^{1/2}$

---

of thousand and the centroidal distance is in the range of hundred. It implies that the area feature can dominate on the centroidal distance. As a result, other features become insignificant to determine the identity of a person. Therefore, all the features $f_i \in F$ are normalized to $\hat{f}_i \rightarrow \{0, 1\}$. The features are normalized as

$$\hat{f}_{i,j} = \left(f_{i,j} - \min(f_i)\right) / \left(\max(f_i) - \min(f_i)\right) \tag{3}$$

where $f_{i,j}$ represents the ith feature of the jth subject, $max(f_i)$ and $min(f_i)$ represent the maximum and minimum values of the ith feature in F, respectively. Now, the feature set $\hat{F} = \left\{\hat{f}_i\right\}_{i=1}^{16}$ is experimented.

4 Experimental Description

The experimental methods are twofold. In the first case, the $HANDCAPTCHA$ images are solved to discriminate whether the user is a human or an automated system. In the next experiment, authentication of the user is performed.

(a) $HANDCAPTCHA$ generation dataset

The $HANDCAPTCHA$ creation method requires three different sets of images.

(i) About 400 I_{BG} images are collected from the Internet. It includes various types of high-quality images such as of texture, natural scenes, flowers, art, and other objects.

(ii) The genuine hand image dataset D_{GH} includes 200 subjects with two hand images per subject of the Bosphorus hand database. The genuine dataset contains the hand images of 100 left-hand and 100 right-hand subjects. Two images of the same hand are called genuine because both of the images are captured using the same sensor device and the properties of the images like image dimension, background variations are similar. Therefore, the entropies of I_{GH} images are similar.

(iii) Altogether 350 fake hand images with different characteristics are collected from the Internet and social media with sufficient variations. The fake hand dataset D_{FH} includes the images of celebrity, colorful design printed on hand, prosthetic hand, dorsal hand, carton, emoticons, hand with gloves, baby handprint, 3-D and infrared hand images, robotic arm, and stylish hands of women with different design and tattoos, and conventional 2-D fake hand images. These are called fake images because the properties of every image are different. Therefore, the entropies of I_{FH} images are dissimilar, and variations of entropy are significant.

Utmost nine hand images are considered in every $HANDCAPTCHA$ image. Two genuine images of the same person and five to seven fake images are selected randomly. At first, the volunteers are trained how to solve the $HANDCAPTCHA$ problem. For this purpose, mainly ten undergraduate student volunteers are selected and trained how to choose the correct I_{GH} images and how to label according to their positions in the I_{HC} image. Based on the responses and comments, the $HANDCAPTCHA$ generation algorithm has been modified accordingly. For example, variation between the background and foreground images should be high. In some challenging cases, horizontal or vertical stripes in I_{HC} can cause visualization problem because the I_{FH} images are also very identical to the I_{GH} images. As a result, those operations are not tried out further and neglected. Initially, during learning phase to solve the first $HANDCAPTCHA$, a bit of more time typically within 3–7 s is required. Afterward, the users can solve the problem within 2 s. In the training phase, more than 50 numbers of I_{HC} images are given to the volunteer users to solve. The performance of the $HANDCAPTCHA$ is calculated regarding the accuracy to solve it correctly. The accuracy of the $HANDCAPTCHA$ is based on human responses and is defined as

Table 1 Performance evaluation of $HANDCAPTCHA$ based on human responses

Phase	Unique I_{HC}	Users	Responses per user	Total responses	Accuracy (%)
Training	20	10	5	50	94
Testing	300	100	6	600	98.34

$$Accuracy(\%) = (total\ number\ of\ correct\ responses\ /\ total\ number\ of\ responses) \times 100 \tag{4}$$

Based on the responses, the accuracy is found as 94% in the training phase.

During the testing phase, 300 unique I_{HC} images are created. Responses from 100 undergraduate students are collected by the trained volunteers. Each student member is given a set of six unique I_{HC} images to solve. Every set includes at least one I_{HC} image with $N_{FH} = 5$, 6, and 7. Therefore, altogether 600 responses are collected, based on which the accuracy is calculated.

The users have correctly found out the locations of the two I_{GH} images, not only by the similarity in the hand structure, size, sex, skin color, and left or right hand but also followed the resemblance in the patterns of the major palm lines. Any special hand accessories like ring or bracelet are also an important factor to identify the I_{GH} images. This visualization and perception capability of the human eye provides the sensitivity to correctly recognize the genuine hands which are not easily possible with an automated script.

Ten responses are found wrong out of total 600 responses. The main reason is identifying the positions incorrectly where the I_{GH} images are located. It happens mainly when one or two positions are blank, i.e., $N_{FH} < 7$. One more reason is the high value of alpha-blending and opacity that causes difficulties in the visualization of the user. In three cases during the $HANDCAPTCHA$ generation, noise has been associated which causes an abrupt change in the intensities of the pixels which results in the change of color in the palm and boundary region. Moreover, it can also damage the I_{GH} image in I_{HC}. Hence, during preprocessing the clarity of I_{GH} images should not be lowered or damaged (Table 1).

(b) Probability estimation of a random guess and attack

Let the size (S_{HC}) of I_{HC} image is $X \times Y$ pixels, and size (S_{GH}) of every I_{GH} image is $m \times n$ pixels. The probability of selecting any one of two I_{GH} images by random guess is $p_1 = 2 \times S_{GH} / S_{HC}$. Now, the second I_{GH} should be chosen from remaining pixels $S_U = S_{HC} - S_{GH} = (X \times Y - m \times n)$ pixels. Next, the probability of selecting the other I_{GH} image is $p_2 = S_{GH}/S_U$. Therefore, the total probability of solving the $HANDCAPTCHA$ image by random guess by a user or program is given as $P = p_1 \times p_2$. Here, S_{HC} is fixed to 460×460 pixels, and the S_{GH} has been varied from 100×100 to 150×150 pixels. Thus, the minimum and maximum probabilities of a solution by arbitrary responses are $P_{min} = 0.004688$ and $P_{max} = 0.025304$, respectively. It implies that a given I_{HC} with a fixed higher dimension and the I_{GH} images with a lower dimension provide better security regarding lower

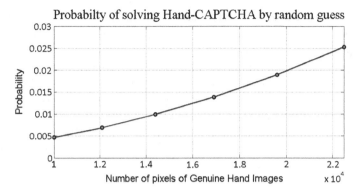

Fig. 6 Probability estimation with different genuine hand image dimension

probability to solve the *HANDCAPTCHA* by an attack with a random guess. However, during execution, the dimensions of both I_{GH} images may differ. Therefore, for any practical situation, the probability P_c satisfies $P_{min} \leq P_c \leq P_{max}$. The *HANDCAPTCHA* generation algorithm is dynamic and scalable in all aspects. Different types of images with various dimensions can be included in the fake dataset, and other transformations and distortion methods can be incorporated. The probability will increase if the dimension of I_{GH} increases and it is evident in Fig. 6.

(c) **Hand biometric verification**

Verification experiments are conducted on the *Bosphorus* hand database, created at the *BOGAZICI University* [4]. Three images per hand at various sessions have been acquired using a low-resolution HP Scanjet flatbed scanner with 383×526 pixels at 45 dpi. Description of the database is mentioned in [4]. The left-hand images of 200 subjects are experimented to assess the performance and reliability of the present method. Based on the three hand images, two images are selected for training, and remaining one image is used for testing. Identification experiments are performed using the kNN classifier with $k = 3$. The kNN classifier is commonly used in pattern recognition due to its algorithmic simplicity and lower time complexity. In the verification, the performance is evaluated regarding the equal error rate (EER). A test feature set has been compared to his/her stored templates on a given distance threshold. Distances between a claimer and the enrolled feature matrix are calculated. If the differences are within the threshold, then a person is accepted as genuine, otherwise rejected as an imposter. The distance threshold (*th*) is defined as

$$\text{th}_{e,g} = \sum_{i=1}^{16} \left[abs\left(\hat{\text{f}}_{e,i} - \text{b}_{g,i}\right) \Big/ mean\left(\hat{\text{f}}_i\right) \right] \tag{5}$$

where $\hat{\text{f}}_{e,i}$ represents the ith feature of the e registered user, and $\text{b}_{g,i}$ means the ith test feature of a claimant g. The mean value of the ith feature is calculated over the

Table 2 Performance evaluation of hand geometric authentication

Subjects (%)	100	200
kNN accuracy	100	99.5
EER	3.50	3.93

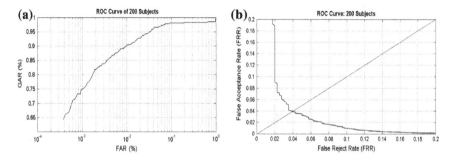

Fig. 7 ROC curve **a** the FAR versus GAR represented in logarithmic scale; **b** the FRR versus FAR to estimate the EER through the diagonal line

training dataset. Different parameters for verification with 200 genuine subjects are given as

Training set $= 2 \times 200$; test set $= 200$; total comparison $= 400 \times 200$; genuine comparison $= 400$; and imposter comparison $= 79600$ (Table 2).

The receiver operating characteristic (ROC) curve is plotted in Fig. 7 to justify the genuine acceptance rate (GAR) with the false acceptance rate (FAR).

(d) **Performance comparison**

The probability estimation for a random guess attack in face-CAPTCHA [15] to locate the genuine face images with size 100×100 pixels is 0.00688, and the accuracy to solve by the human is 98%. In contrast, the probability of random guess attack in *HANDCAPTCHA* for the same image dimension is 0.004688, and the accuracy of solving by a person is 98.34%. Similarly, the probability of random guess attack considering two genuine face images in face-DCAPTCHA [16] with dimension of 400×300 pixels and tolerance of 80×80 pixels is 0.00598. On the other hand, the probability of the proposed work for the same experimental constraint is 0.00188. Therefore, the proposed system is lesser vulnerable to attack.

On the other hand, a quantitative performance comparison with few established contact-free hand biometric systems over the Bosphorus hand database is presented in Table 3. However, the preprocessing technique, feature set definition with extraction method, and other experimental setup are dissimilar in different works. The proposed method is comparable to the approaches based on the same database, mentioned in Table 3.

Table 3 Performance comparison of hand biometrics

Author	Approach	Subjects	Identification (%)	Verification (%)
Yörük et al. [9]	ICA	100	98.81	GAR: 98.21
Dutağaci et al. [4]	Geometric with LDA	200	98.22	Not reported
Proposed	Geometric	200	99.5	EER: 3.93

5 Conclusion

A new approach for hand biometric recognition with $HANDCAPTCHA$ verification method is presented. The image-based $HANDCAPTCHA$ method is competitive over the traditional text-based and face image-based CAPTCHA. This system is also advantageous over the password-based authentication scheme and malicious attacks. The probability of random guess attack by an automated script is encouraging in this work. The main advantage of $HANDCAPTCHA$ is to enhance the security and reliability of hand biometric system in a challenging environment. The scalability of $HANDCAPTCHA$ algorithm offers its suitability to be incorporated in Web-based applications and other handheld smaller devices like a smartphone. As a future enhancement, responses from more people should be collected to measure its accuracy. Moreover, another automated algorithm should be implemented to verify its performance and robustness against different spam attacks. In addition, the present hand biometric verification should be tested with new features and more subjects to enhance its potential and applicability in a high-security environment.

All the datasets used in this paper all are freely available from the Internet. The Bosphorus hand database is available freely for research, and has been collected from Prof. B. Sankur with maintaining the ethics.

References

1. Jain, A.K., Ross, A., Prabhakar, S.: An introduction to biometric recognition. IEEE Trans. Circuits Syst. Video Technol. **14**(1), 4–20 (2004)
2. Jain, A.K., Ross, A., Pankanti, S.: Biometrics: a tool for information security. IEEE Trans. Inf. Forensics Secur. **1**(2), 125–143 (2006)
3. Böhme, R., Freiling, F.C., Gloe, T., Kirchner, M.: Multimedia forensics is not computer forensics. In: IWCF 2009. LNCS, vol. 5718, pp. 90–103 (2009)
4. Dutağaci, H., Sankur, B., Yörük, E.: A comparative analysis of global hand appearance-based person recognition. J. Electron. Imaging **17**(1), 011018-1–011018-19 (2008)
5. Galbally, J., Marcel, S., Fierrez, J.: Image quality assessment for fake biometric detection: application to iris, fingerprint, and face recognition. IEEE Trans. Image Process. **23**(2), 710–724 (2014)
6. Marsico, M.D., Marchionni, L., Novelli, A., Oertel, M.: FATCHA: biometrics lends tools for CAPTCHAs. Multimed. Tools Appl. (2016). https://doi.org/10.1007/s11042-016-3518-8
7. Duta, N.: A survey of biometric technology based on hand shape. Pattern Recogn. **42**(11), 2797–2806 (2009)

8. Reillo, R.S., Avila, C.S., Macros, A.G.: Biometric identification through hand geometry measurements. IEEE Trans. Pattern Anal. Mach. Intell. **22**(10), 1168–1171 (2000)
9. Yörük, E., Konukoğlu, E., Sankur, B., Darbon, J.: Shape-based hand recognition. IEEE Trans. Image Process. **15**(7), 1803–1815 (2006)
10. El-Alfy, E.S.M.: Automatic identification based on hand geometry and probabilistic neural networks. In: 5th IEEE International Conference on New Technologies, Mobility and Security (NTMS), pp. 1–5 (2012)
11. Luque-Baena, R.M., Elizondo, D., López-Rubio, E., Palomo, E.J., Watson, T.: Assessment of geometric features for individual identification and verification in biometric hand systems. Expert Syst. Appl. **40**(9), 3580–3594 (2013)
12. Charfi, N., Trichili, H., Alimi, A.M., Solaiman, B.: Novel hand biometric system using invariant descriptors. In: IEEE International Conference on Soft Computing and Pattern Recognition, pp. 261–266 (2014)
13. Kanhangad, V., Kumar, A., Zhang, D.: A unified framework for contactless hand verification. IEEE Trans. Inf. Forensics Secur. **6**(3), 1014–1027 (2011)
14. Wang, M.H., Chung, Y.K.: Applications of thermal image and extension theory to biometric personal recognition. Expert Syst. Appl. **39**(8), 7132–7137 (2012)
15. Goswami, G., Singh, R., Vatsa, M., Powell, B., Noore, A. Face recognition CAPTCHA. In: 5th IEEE International Conference on Biometrics: Theory, Applications and System (BTAS), pp. 412–417 (2012)
16. Goswami, G., Powell, B.M., Vatsa, M., Singh, R., Noore, A.: FaceDCAPTCHA: face detection based color image CAPTCHA. Future Gener. Comput. Syst. **31**, 59–68 (2014)
17. Baecher, P., Buscher, N., Fischlin, M., Milde, B.: Breaking reCAPTCHA: a holistic approach via shape recognition. Future Chall. Secur. Priv. Acad. Ind. **354**, 56–67 (2011)
18. Von Ahn, L., Maurer, B., McMillen, C., Abraham, D., Blum, M.: Recaptcha: human-based character recognition via web security measures. Science **321**(5895), 1465–1468 (2008)
19. Otsu, N.: A threshold selection method from gray-level histograms. IEEE Trans. SMC **9**(1), 62–66 (1979)

New Approach to Smartwatch in Human Recognition

Paweł Kobojek, Albert Wolant and Khalid Saeed

Abstract Nowadays, personal IoT devices and wearable electronics are taking international markets by storm. Each of these devices is equipped with a variety of different sensors. Throughout the wide range of specific models, one of these sensors is accelerometer. Presented work aims to study the possibility of employing onboard accelerometers of smartwatches to perform owners recognition. We introduce a way of checking the time as a behavioral biometric feature. As a part of this effort, dedicated dataset was created. Then classification algorithms were adjusted and tested. Additionally, comparison was done, between above-mentioned method and more traditional approach, which included computer vision.

1 Introduction

In modern world, the way people interact with their environment is mostly dominated by usage of computers and personal electronic devices. Users spend much of their time operating them and generate amounts of data never seen in the history before. Such data-rich and dense world is a great place to introduce and study behavioral biometric features.

During last years, we have witnessed wearable electronics transition from being prototypes and concepts to becoming everyday items with high accessibility. Especially, smartwatches and smartbands became very popular. Because of that, we decided to study possible ways of harvesting the data generated by those devices in task of user verification. As mentioned devices are wrist operated, we are focusing on the most natural way they are used, namely checking the time or more generally checking the smartwatch screen.

P. Kobojek · A. Wolant · K. Saeed (✉)
Faculty of Computer Science, Bialystok University of Technology, Bialystok, Poland
e-mail: k.saeed@pb.edu.pl

© Springer Nature Singapore Pte Ltd. 2018
R. Chaki et al. (eds.), *Advanced Computing and Systems for Security*,
Advances in Intelligent Systems and Computing 666,
https://doi.org/10.1007/978-981-10-8180-4_2

Successful method of verification would make real-life applications possible, such as unlocking smartwatch screen, only when it is looked at by the owner. Since more traditional verification methods, like PIN numbers, are very inconvenient for smartwtach users, we believe that our work might significantly impact the way these devices are used.

2   Related Work

Significant effort has been made in the field of behavioral biometrics throughout the years. Below we present only handful of research papers published and discuss how they correspond to our work.

In [1], authors proposed criteria of a good behavioral biometrics. These criteria are as follows: universality, uniqueness, permanence, collectability, performance, acceptability, circumvention. Let us discuss how our work relate to those criteria.

Obviously, our method is relevant only to smartwatch owners. In this group, checking the screen of the watch is highly universal and easy to collect. Since measurement happens without any special input from the user and collected data is not very sensitive, we cannot see any problems regarding acceptability. As for now, any kind of circumvention also seems unlikely, since it would require moving smartwatch exactly in the same way as the owner. Although it might be possible to do that, based on stolen information of previous users movements, providing that this sensitive data is protected, randomly getting accurate movement should be difficult. The aim of this work is to determine uniqueness and performance of this behavioral biometric feature. Question about permanence stays open. On one side, gesture of checking the time is well trained and intuitively known for each individual. On the other side, it might change due to aging and other external factors, such as different weight of new watch, intensive upper body training, or broken arm bones.

In [2], authors took upon themselves task of comprehensive summary of modern behavioral biometrics. They present results from multitude of different sources and concerning variety of behavioral features, from ways that users interact with computers on multiple levels, to more natural ones, like gait. They also created taxonomy of features categorizing them into five classes as follows: authorship-based biometrics, human–computer interaction (HCI)-based biometrics, indirect HCI-based biometrics, motor-skills biometrics, and purely behavioral biometrics. In this system, the presented method should be classified as a purely behavioral biometrics alongside the way person walks, types or grasps a tool.

Topic of purely behavioral biometric features is one of the most exciting and explored avenues of research these days. For example in [3], authors used a tailored version of kNN algorithm to tackle keystroke dynamics problem. As our solutions also use kNN, we found their work helpful.

The idea of using accelerometer in biometrics has already been tried in a few different applications. One of them was described in [4]. Authors used accelerometer data to distinguish between activities such as walking, vacuuming, or doing sit-ups. Results they presented showed that their methods yield satisfying results.

As smartwatches became popular, the idea of employing them for biometrics usage emerged in research as well. In [5], authors studied possibility of authenticating users based on accelerometer and gyroscope data. Their premise was to analyze the movement of people arms as they walk and used this data to distinguish between users. They compared multiple methods of classification and concluded that it is feasible approach.

Other teams tried using hand gestures as source of unique movement tracking data. In [6], authors proposed algorithm in which each user picks up his or her own gesture, then practices it multiple times, and then is tracked while doing it. Although the idea seems interesting and results were promising, the notion of creating own gestures makes it less convenient for the users. Other works, like [7–9] emphasized the concept of predefined gestures for all the users.

3 Data Acquisition

As presented concept is quite novel and no existing sets of example data were available to us, first part of the project was to collect proper data.

Device used to gather it was first-generation Apple Watch. Since it is one of the most popular devices in personal IoT segment, the assumption was it will represent the category well, in terms of possible onboard sensors and their performance.

Each of the 36 participants of the study was asked to check the time on provided watch several times. Person started gesture on sign given by the operator conducting the experiment. When subject felt satisfied with the checking, signaled the operator to stop capturing process. This way, we gathered data of only first phase of the movement. It is critical for the seamless use of finished product, since the verification suppose to unlock the screen of the smartwatch. As observed, the average time of checking the time was around 2 s. During this period, around 20 data points were collected.

Apple Watch is equipped with multiple sensors. For purpose of this work, we collected data from several of them. Firstly, data from accelerometer was recorded. It shows proper acceleration values in each of the three axes. This gives general idea of how the watch was moved during the experiment. It would be possible to build models of watch movement based on such data and might be fruitful avenue of future research.

Apple Watch API makes it easy to use data from other sensors as well. Device, beyond accelerometer, has onboard gyroscope, for orientation measurement, and magnetometer, for measurement of direction and strength of Earth's magnetic field. Data from all of those is combined into easy to use structures.

Attitude represents the orientation of a body relative to a given frame of reference. Rotation rate structure contains data specifying the device's rate of rotation around the axis. It is constructed based on the gyroscope measurements. Gravity shows gravity acceleration expressed in the device's reference frame. Magnetic field structure yields the magnetic field vector with respect to the device.

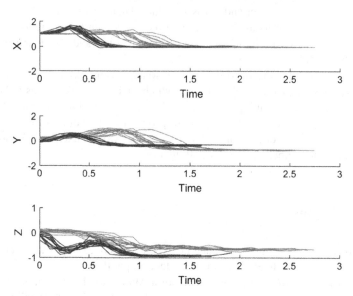

Fig. 1 Example of accelerometer data. Each plot shows measurements in different axes. Samples of the same color are collected from the same person

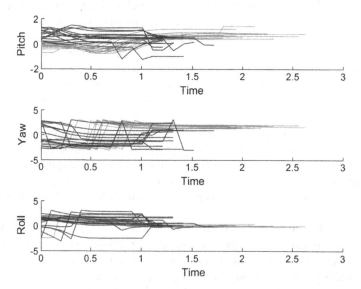

Fig. 2 Example of attitude data. Each plot shows measurements in different dimension. Samples of the same color are collected from the same person

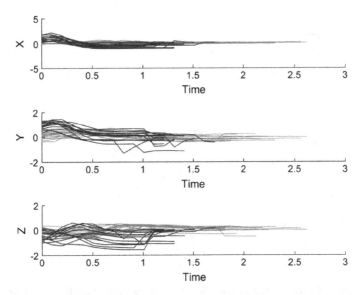

Fig. 3 Example of user relative acceleration data. Each plot shows measurements in different axes. Samples of the same color are collected from the same person

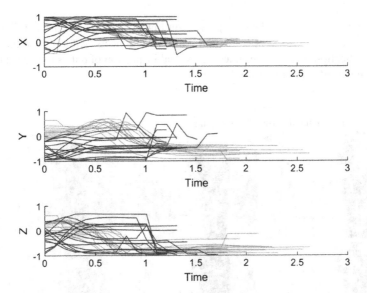

Fig. 4 Example of gravity data. Each plot shows measurements in different axes. Samples of the same color are collected from the same person

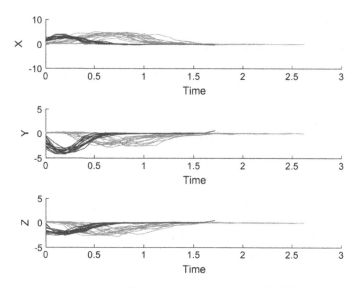

Fig. 5 Example of rotation rate data. Each plot shows measurements in different axes. Samples of the same color are collected from the same person

Examples of collected data are present on Figs. 1, 2, 3, 4, and 5. Each line represents data series. Different colors mark different subjects. Although on the stage of data collection it cannot be stated yet, visual inspection of data showed promises that proposed idea might be viable.

Table 1 Frames captured during data gathering

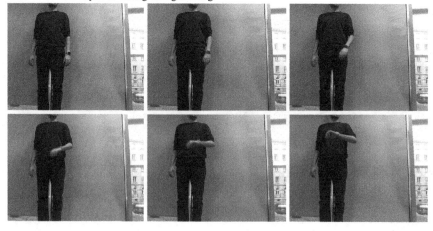

While recording each data point, subjects were recorded using standard 1080 p camera. Example of captured frames is shown in Table 1. People were facing camera directly, as shown in the pictures.

4 Previous Algorithm

Feature selection and extraction

Raw accelerometer and Device Motion data are the basis of our method. However, not everything which is gathered by those sensors is a viable metric for our system. For the most obvious example, magnetic field three-dimensional vector (x, y, z) values were always equal to 0 across whole dataset, so it was first to be removed. By plotting the graphs depicting how each value changes with time, one can spot that some of them are not viable to use for further classification.

Raw data with selected features were then subjected to feature extraction procedure. As it was stated before, for each subject, we have collected over a dozen of samples. Each such sample is a sequence of accelerometer or multiple Device Motion recordings. In order to create feature vectors out of this data, we have employed a simple and already used in classification based on accelerometer data technique [4]. For each acceleration coordinate, as well as for each Device Motion value, we have computed a mean and standard deviation from every corresponding value of the sample sequence. In case of accelerometer, which has three coordinates, it resulted in six values per sample, and when it comes to device motion data, which has total of 13 different values it resulted in 26 dimensional vector.

Preliminary results

In order to test our hypothesis that smartwatch sensors are viable for user verification, we have employed a simple k- nearest neighbors (kNN) algorithm for multiclass classification. We have tested multiple parameters—13 different distance measures and 4 k values, which means total of 52 settings. What is more, we performed tests using different sets of features. First test was performed using only accelerometer data, second involved only device motion data, and third test used all the data combined. After cleaning of the data, total dataset consisted of 430 samples from 36 subjects. Each subject in this experiment was assigned to a separate class. The dataset was split into train and test set (80% and 20% of original data size, respectively). It is worth noting that subjects do not have an equal count of samples. Selected results are shown in Table 2. The numbers in bold are best accuracy for the subset.

Table 2 Accuracy values for selected kNN parameters on different subsets of gathered data

(k, distance)	Accelerometer (%)	Device motion (%)	Both (%)
(1, max)	**71**	23	45
(1, braycurtis)	64	**56**	57
(1, manhattan)	65	55	**60**

The distance functions shown in the table are computed according to Eqs. 1, 2, and 3. x and y are the two vectors between which we compute the distance.

$$D = \max_i(|x_i - y_i|) \tag{1}$$

$$D = \frac{\sum_i |x_i - y_i|}{\sum_i |x_i + y_i|} \tag{2}$$

$$D = \sum_i |x_i - y_i| \tag{3}$$

The resulting accuracy achieved by a very simple algorithm (kNN) gives a premise that there truly is a significant signal in the data we gathered. It can be seen that using only accelerometer data results in better accuracy. This is because Device Motion data contains values which are similar between subjects. Either better feature selection/extraction or using more sophisticated classifiers would probably improve this.

5 The Modified Algorithm

The previous algorithm used Device Motion and/or accelerometer data and simple kNN as a classifier. In this section, we will describe a similar algorithm which uses a more powerful model—support vector machines. Also, we have tested a kNN classifier on video data which was gathered as well.

Support Vector Machine as the classifier

Support vector machines [10] are widely used and well-known classifiers. We have used them as a replacement for less sophisticated k Nearest Neighbors. Results of applying SVMs on our dataset are shown in Table 3. The table shows results for four different SVM kernels. As it was expected, the accuracies are significantly higher than corresponding ones for kNN. The best we managed to achieve was the result of fourth degree polynomial kernel.

Table 3 Accuracy values for selected SVM kernels on different subsets of gathered data

kernel	Accelerometer (%)	Device motion (%)	Both (%)
rbf	100	91	92
sigmoid	100	97	97
poly (3th deg.)	100	98	98
poly (4th deg.)	100	98	100

Table 4 Accuracy values for best performing kNN settings on video frames

(k, distance)	Accelerometer (%)
(1, manhattan)	99
(1, braycurtis)	99
(1, canberra)	99
(1, cosine)	98

Usage of videos

As a measure of comparison, we also took videos of subjects while they were check-ing the time. Then we used motion tracking algorithms to construct model of each person moving and check how the accuracy of verification obtained this way will be comparable to the accuracy of method presented above. For motion tracking, we used standard approach of calculating optical flow for each frame. As a base of the model, we used generated focal points. The 2D points were then flattened to 1D vector of x-coordinate and y-coordinate alternately and padded with zeros to have equal length. Such vectors were input to the same kNN algorithm as before. Table 4 shows best achieved accuracies. The dataset was again split into train and test set (80% and 20% of original data size, respectively).

Distance functions shown in Table 4 are computed according to Eqs. 2 (braycurtis), 3 (manhattan), 4, and 5. $x \cdot y$ is a dot product of x and y. $||x||_2$ is a second norm of x.

$$D = 1 - \frac{x \cdot y}{||x||_2 ||y||_2} \tag{4}$$

$$D = \sum_i \frac{|x_i - y_i|}{|x_i| + |y_i|} \tag{5}$$

6 Conclusions

In this paper, we introduced behavioral biometric feature and layed the groundwork for its further research. We created and presented dataset gathered to study the way of checking the time on smartwatch as biometric feature. Using data analysis tools we showed that differences between people can be observed and measured and are distinctive enough for classification. We have shown that the way a human checks the time on a watch can be a novel and viable behavioral biometric feature. It is also worth noting that the valuable signal is seen despite relatively low frequency (10 Hz) of gathered data and using the simplest methods of feature extraction. Furthermore, we have tested an application of popular SVM classifiers for this problem which resulted in very high (more than 90%) accuracy, which is significantly higher than k Nearest Neighbors. SVM with 4 h degree polynomial kernel classification accuracy was especially high—it has achieved 100% on accelerometer only data and on accelerometer and Device Motion data combined. However, a reader should be aware that this high performance is achieved on a relatively small dataset. We also showed that using videos of a subject, checking the time is also a useful feature for recognition. Accuracies were high here as well, but size of the dataset was also relatively small.

7 Further Work

We are currently investigating the possibility of creating a tailored algorithm to perform user verification based on data presented in above-mentioned dataset. Also, we are assessing which parameters of measured motion are the most important ones in regard to applications in biometric. Since data from camera turns out to be a useful metric as well, we are working on merging both accelerometer and/or Device Motion data with signal from camera. As it was described in Sect. 3, video was recorded simultaneously with accelerometer and Device Motion data gathering. Thus, both these inputs may be easily combined to form a promising description of the recognized user. We also want to extend our dataset to have a better insight on classifiers performance. Another important plan for future is to employ an online verification of a user based on data we have gathered. This could be useful for real- life applications, i.e., ongoing verification in which the user is prompted for PIN if an algorithm decides the way he or she checks time is not matching the pattern.

The authors declare that both the subject and the idea of its realization are new and the dataset is ours. Nothing was taken from any other reference.

Acknowledgements This work was supported by Białystok University of Technology and Warsaw University of Technology.

References

1. Jain, A.K., Ross, A., Prabhakar, S.: An introduction to biometric recognition. IEEE Trans. Circ. Syst. Video Technol. **14**, 4–20 (2004)
2. Yampolskiy, R.V., Govindaraju, V.: Behavioural biometrics: a survey and classification. Int. J. Biometrics **1**(1), 81–113 (2008)
3. Panasiuk, P., Saeed, K.: A Modifed Algorithm for User Identification by his Typing on the Keyboard. Advances in Intelligent and Soft Computing, vol. 84. Image Processing and Communications Challenges **2**, 113–120 (2010)
4. Ravi, N., Dandekar, N., Mysore, P., Littman, M.L.: Activity recognition from accelerometer data. Am. Association Artificial Intell. (2005)
5. Kumar, R., Phoha, V.V., Rahul, R.: Authenticating users through their arm movement patterns. arXiv:1603.02211 [cs.CV]
6. Lewis, A., Li, Y., Xie, M.: Real time motion-based authentication for smartwatch. In: 2016 IEEE Conference on Communications and Network Security (CNS): IEEE CNS 2016 - Posters (2016)
7. Saravanan, P., Clarke, S., Chau, D.H.P., Zha, H.: Latentgesture: active user authentication through background touch analysis. In Proceedings of 2nd International Symposium of Chinese CHI, pp. 110–113 (2014)
8. Shahzad, M., Liu, A.X., Samuel, A.: Secure unlocking of mobile touch screen devices by simple gestures: you can see it but you can not do it. In: Proceedings of MobiCom '13, pp. 39–50 (2013)
9. Yang, J., Li, Y., Xie, M.: Motionauth: motion-based authentication for wrist worn smart devices. In: Proceedings of the 1st Workshop on Sensing Systems and Applications Using Wrist Worn Smart Devices, pp. 550–555 (2015)
10. Gunn, S.R.: Support Vector Machines for classification and regression. Technical report, University of Southampton, Faculty of Engineering, Science and Mathematics, School of Electronics and Computer Science (1998)
11. Adamski, M., Saeed, K.: Signature verification using contextual information enhancement and dynamic programming. J. Med. Inf. Technol. **12**, 35–40 (2008)
12. Nassi, B., Levy, A., Elovici, Y., Shmueli, E.: Handwritten signature verification using hand-worn devices. arXiv preprint arXiv:1612.06305 [cs.CR]
13. Xu, C., Pathak, P.H., Mohapatra, P.: Finger-writing with smartwatch: a case for finger and hand gesture recognition using smartwatch. In: Proceedings of the 16th International Workshop on Mobile Computing Systems and Applications. ACM, pp. 9–14 (2015)

Retina Tomography and Optical Coherence Tomography in Eye Diagnostic System

Maciej Szymkowski, Emil Saeed and Khalid Saeed

Abstract Eye diagnostic, two-step method based on retina color image and Optical Coherence Tomography is presented in this paper. A new robust algorithm, by which various eye diseases can be diagnosed, was implemented as an essential part of the work. The approach comprises two steps. The first one deals with the analysis of retina color image. In this stage, an algorithm was implemented to especially search hard exudates. If the algorithm returns positive, it means at least one hard exudate was found. Moreover, it may return hesitant results in the case of changes that look like hard exudates. During the second step, additional analysis of Optical Coherence Tomography image is done. In this stage, the algorithm is looking for confirmation of hard exudates, which were found during the first step. The authors' approach gives more confidence in the cases of small exudates or initial stages for eye diseases.

Keywords Image analysis · Retina in slit lamp examination · Optical Coherence Tomography · Eye diseases · Automated diagnosis

1 Introduction

Today, hard exudates are one of the most common pathology changes in the course of different eye diseases. For instance, these changes could appear as a part of diabetic retinopathy. Moreover, if they are detected at the proper time, treatment to prevent macular edema and advanced stage of diabetic retinopathy can be applied.

M. Szymkowski (✉) · K. Saeed
Faculty of Computer Science, Bialystok University of Technology, Bialystok, Poland
e-mail: m.szymkowski@student.pb.edu.pl

K. Saeed
e-mail: k.saeed@pb.edu.pl

E. Saeed
Department of Ophthalmology Faculty of Medicine, Medical University of Bialystok, Bialystok, Poland
e-mail: emilsaeed1986@gmail.com

© Springer Nature Singapore Pte Ltd. 2018
R. Chaki et al. (eds.), *Advanced Computing and Systems for Security*,
Advances in Intelligent Systems and Computing 666,
https://doi.org/10.1007/978-981-10-8180-4_3

This remedy is essential for patients to preserve their vision. Ophthalmologists can detect these changes by the usage of retina in slit lamp examination or Optical Coherence Tomography. A few interesting approaches to retina and OCT processing are introduced here. At the beginning, the thought-provoking idea of Eadgahi and Pourezza [1] is described. Authors of this publication have prepared a method based on morphological operations on a retina color image. As the first step, vessel elimination and extraction of bright components were done. Both aims are obtained with the usage of a few morphological operations, like closing or opening, and image subtraction. The second step of this algorithm is to detect the optical disk on retina image. It is done by usage of entropy calculation. After this operation, the next step is performed. It uses Sobel operator by which all edges are detected. Detection of exudates is done as the last step after edge detection. Authors for this paper claimed their approach gave about 78% of average sensitivity on exudates.

In [2], an interesting approach that consists of four steps was presented. At the beginning, authors proposed image preprocessing. This step is based on filtering operations, like Gaussian Filter or Median Filter, and changing image colors with the usage of only green channel values. Then fovea localization is mentioned as the second step. In this case, authors used a few geometric transformations to perform this operation. As the third one, detection of hard exudates is proposed. This idea is based on image segmentation and adaptive thresholding. At the end of the whole procedure, the authors proceed with the detection of exudates by classifying with a medical polar coordinate system. The authors claim this approach gave a recognition rate of about 98%.

Another method within the state of the art is what the authors of [3] had worked out. They presented a simple method that gave interesting results. At the beginning, they use Contrast Limited Adaptive Histogram Equalization method. This step is done due to the fact that it increases the contrast between exudates and other parts of the retina. Then, the segmentation takes place. In this case, authors used Contextual Clustering to divide image into two parts—background and alternative class in which exudates could be visible. As the last step, feature vector generation is created. It is based on those features that describe all of the regions that may probably show the possibility of exudates existence.

An interesting approach was given in [4]. The algorithm that was described in this paper consists of seven steps in which the authors make simple image operations as changing all pixel channels to green channel value or color normalization and Haar wavelet decomposition and reconstruction. The authors of this paper mentioned that they used four databases of fundus images. Their approach gives no more than 22.48% of accuracy level. They also claimed that this paper was the initial step of their work under hard exudate detection and it could be used by ophthalmologists only in early detection of non-proliferative diabetic retinopathy.

Authors of [5] also take into consideration diabetic retinopathy as the disease that has to be detected. At the beginning of their algorithm, also extraction of green channel is done and then morphological dilation is used. As another step, clustering was mentioned. Within this step, the authors use two different algorithms: Linde–Buzo–Gray and k-means algorithms. On the basis of the obtained results, one clustering algorithm is selected. At the end of the whole method, post-processing is applied. This step consists of the detection and elimination of the optical disk. Authors claim that their database consists of 89 pictures. Moreover, in the work, the accuracy of the proposed approach is no greater than 76%.

In the case of OCT processing, there are not as much approaches as it is in the case of retina in slit lamp examination. In [6], authors prepared a comparison between different noise reduction algorithms that could be used in OCT images. Tested algorithms included different filtering methods, anisotropic diffusion, soft wavelet thresholding and multiframe wavelet thresholding. Precision of the denoising process was evaluated on the basis of automated retina layer segmentation results. Experiments were conducted with a set of 10 healthy scans and 10 samples with vitreoretinal pathologies. Authors claimed that anisotropic diffusion and wavelet thresholding give the best results and allow for better retina segmentation for both of sets. Interesting results were obtained with multiframe wavelet thresholding, but this approach provided significant improvement only for retinas with pathological changes.

Authors of [7] proposed a numerical algorithm based on a small-angle approximation of the radiative transfer equation. This idea was developed to reconstruct scattering characteristics of biological tissues from OCT images. Proposed solution describes biological tissue as a layered random medium with a set of scattering parameters in each layer. This set fully describes Optical Coherence Tomography signal behavior versus probing depth. Reconstruction of scattering parameters is performed by a genetic algorithm that was proposed by the authors. The possibility of estimation of these parameters was also studied for various regimes of OCT signal decay.

In [8], authors take into consideration the sensitivity of Optical Coherence Tomography images. This approach consists of identification of retinal tissue morphology characterized by early neural loss from normal healthy eyes. This process is based on calculation of structural information and fractal dimension. Authors' database consists of 74 healthy eyes and 43 eyes with mild diabetic retinopathy.

Another interesting idea was presented in [9]. In this case, authors prepared only preliminary algorithm which localizes the macular edemas. There was no presentation of any kind of classification methods in that work.

Another interesting approach was presented in [10]. Authors claimed that in the case of AMD or DME, their algorithm gives 100% of correctly identified pathological changes and about 86% in normal subject case. At the beginning, OCT image is denoised with BM3D denoising method. In the paper, it was claimed that this step is done due to the fact that noise always exists on the OCT images. Then, as the second step, retinal curvature flattering is discussed. This operation has to reduce the effects of the perceived retinal curvature during OCT image classification procedure. Next, image is cropped. By this step, the proper region of interest is selected. Then, feature vector is created. In their work, the authors used HOG descriptors. As the last operation, the feature vector classification is applied.

In most of the cases analyzed in [1–10], small hard exudates were not taken into consideration. However, sometimes pathological changes of this size can appear on retina color image or Optical Coherence Tomography images. It is essential to detect them so that the ophthalmologist can treat the patient and protect them against the advanced stage of diabetic retinopathy or vision loss. Due to this problem, the authors of this paper started their research in the direction of finding a solution to this important problem. They started accordingly to make use of some already worked out filters and algorithms to apply on the retina images. The most practical approach seemed to be the work on OCT images. In this paper, a new solution that combines the results from retina in slit lamp examination and Optical Coherence Tomography is introduced and discussed.

2 Proposed Methodology

This section presents the proposed algorithm used for detection of hard exudates on the basis of retina color image and Optical Coherence Tomography image. Samples present hard exudates in images obtained from color retina and Optical Coherence Tomography are shown respectively in Figs. 1 and 2.

All steps of retina processing algorithm are shown in Fig. 3. The figure shows the necessary steps from image acquisition to preprocessing and then morphological filtering to end with the extracted pathological changes.

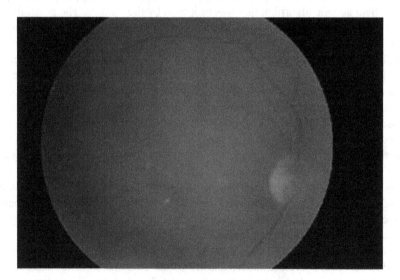

Fig. 1 Retina in slit lamp examination that contains hard exudates

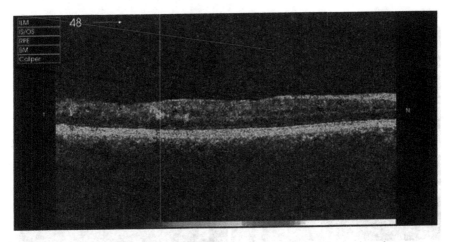

Fig. 2 Image obtained from Optical Coherence Tomography that contains hard exudates

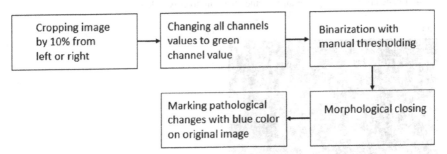

Fig. 3 Block diagram of retina color image processing algorithm

2.1 Retina Color Image Processing Algorithm

In this section, the worked out processing algorithm for retina color image is described. Authors' algorithm starts with cropping input image by 10% from the left or the right sight. The site is chosen on the basis of whether the image presents the left or the right eye. By this step, the optical disk is removed from the under processing image. In the second step, the values of all channels for each pixel are set to the green channel pixel value. One can easily observe that this step changes the image into grayscale. Green channel was chosen due to the fact that only in the case of exudates it has a different value. This means that the rest of the processed retina image has a similar value of green channel. The resulted image after this step is shown in Fig. 4b.

Third and fourth steps are combined with binarization [11] and morphological operation on the image. The first of them was used to change the color of exudates to white. This step was done with manual thresholding. Authors tried different types of binarization methods that were described in [9]. The best results were obtained in the

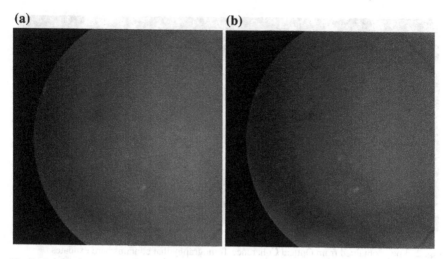

Fig. 4 Image after cropping operation **a** and changing to grayscale **b**

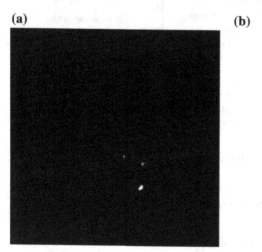

Fig. 5 Image after binarization **a** and after color inversion and morphological closing **b**

case of threshold that was a kind of interval between 80 and 90. After this procedure, a few pixels that were not pathological changes were also marked white. Morphological opening was also used in the processed picture. Before that, the image was inverted so that the white pixels were marked black and vice versa. This operation was done due to the fact that in the case of morphological opening—foreground pixels were marked white and objects pixels were black. The resulting image after all of these operations is shown in Fig. 5.

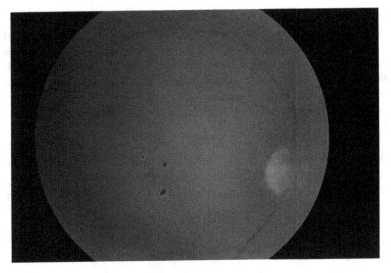

Fig. 6 Pathological changes (hard exudates) detection. The image is enlarged to show the segmented parts clearly

The last step in this procedure showed the pathological changes on the input image. In authors' approach, these changes were marked in blue. All of them are shown in Fig. 6.

2.2 Optical Coherence Tomography Processing Algorithm

In this section, the processing algorithm for Optical Coherence Tomography image is presented. In the case of OCT image processing, the authors used their own approach and their algorithm worked out for this aim.

The main idea of this approach is to detect clearly visible exudates. These changes have a huge influence on human sight and his scope of the visibility. Authors of this idea proposed a method for retinal disorders diagnosing using Optical Coherence Tomography. Sick eyes are picked up automatically by the algorithm. Only significant lesions were taken into consideration. In the case of small pathological changes, the proposed approach has to be modified. Authors prepared a few additional steps for the algorithm that were added because the region of interest in the case of small pathological changes is completely different from the other considered large ones. It has been noticed by the authors that the small changes need a special treatment. And although the obtained results are promising, the authors are still improving their algorithm to detect the smallest possible changes.

The additional modifications were as follows. The first was to leave only red pixels. This is connected with the fact that exudates can be visible in red color on OCT image.

The second was introducing another step, after the morphological closing operation. Authors tried the addition of median filter to remove the unnecessary pixels. After these operations, changes that could be classified as probable pathological changes are surrounded by yellow circles. Now, and after these modifications, the authors' modified algorithm could easily be used to detect hard exudates that are visible in Optical Coherence Tomography image.

The results obtained with the use of the authors' modified approach are shown in Fig. 7.

On the basis of the images shown in Fig. 7, one can easily observe that provided modifications detect not only clearly visible hard exudate but also a few areas that could be classified as "probable" pathological changes. It is significant to detect such very small changes in their initial stages as they can develop into a future pathological change.

3 Comparison with Other Solutions

The authors compared their results with other recently presented solutions of the problem of hard exudates detection in pictures from retina under slit lamp examination.

The first comparison was done with the method in [1]. The results obtained with the usage of both algorithms are shown in Fig. 8.

On the basis of the images shown in Fig. 8, one can conclude that the results are comparable. Both methods can detect hard exudates that are clearly visible. It can also be seen that the exudates are placed in the center of the image. Both methods can detect small pathological changes that can indicate the initial stage of eye disease.

However, authors' method could also detect the small exudates nearby the veins, which was not found by the algorithm of the authors in [1]. From the other side, their algorithm successfully detected exudates nearby optical disk.

The second comparison was done with the method in [4]. The results obtained with the usage of both algorithms are shown in Fig. 9. On the basis of both algorithms, the results presented in Fig. 9 show that the method proposed in [4] does not detect hard exudate (observed in the upper part of Fig. 9a).

Authors' method detected this pathological change. Actually, the authors' algorithm detected more small changes that can probably indicate initial stage of eye disease. The algorithm described in [4] does not detect exudates nearby large lesions either. Only large pathological changes can be found by their algorithm. This might lead to a conclusion that the work in [4] was designed to detect larger exudates.

4 Conclusions

In this work, an approach to combine two methods of hard exudate detection was presented. Only processing algorithms were discussed. The main aim of the study is to work out an algorithm that detects different pathological changes, not only exudates.

(a)

(b)

(c)

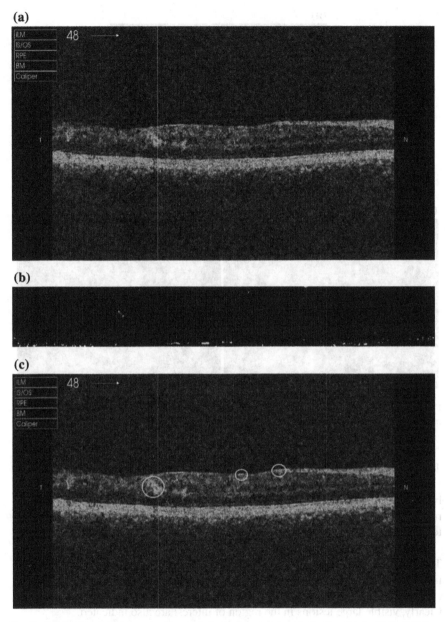

Fig. 7 Original Optical Coherence Tomography image **a** hard exudates detection **b**, Optical Coherence Tomography image with marked probable hard exudates **c**

As was discussed and observed in the case of retina color image, the authors focused on small hard exudates that often cannot be seen by the ophthalmologists.

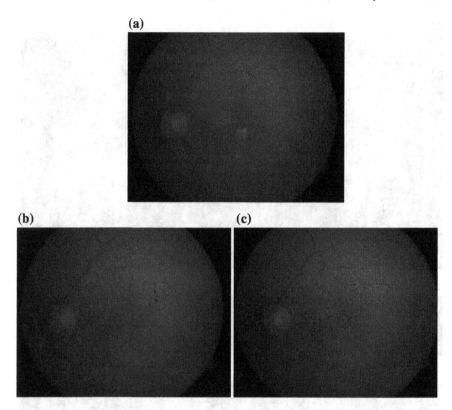

Fig. 8 Original retina image **a** that was used for comparison. Results of segmentation: **b** authors' method, **c** the algorithm described in [1]

The described method can successfully be used to detect some small characteristic changes. The methodology used in the work for Optical Coherence Tomography was based on a solution worked out by the authors that gave satisfying and promising results. The presented modifications can assure hard exudate detection with satisfactory certainty level.

As expected, authors' method does not detect hard exudates nearby optical disk. These pathological changes are not detected because they are not in the region of interest established by the cropping image. However, small lesions and pathological changes nearby the large ones and within the region of interest are correctly detected. Clearly, visible large lesions in the region of interest are also indicated.

The methods described in [1, 4] are more complicated than the algorithm described in this work. However, the results obtained with the algorithm described in [1] and the algorithm presented in this work are close to each other concerning the detection of small pathological changes and large exudates that are in the region of interest established in authors' algorithm. Moreover, in general, such cases, the authors' approach established no worse results than the one in [4]. As a matter of closer

(a)

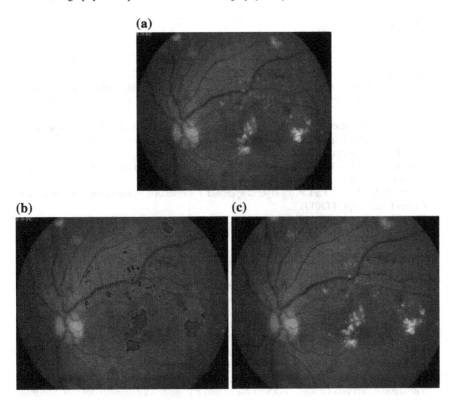

(b)   **(c)**

Fig. 9 Original retina image **a** that was used for comparison. Results of segmentation: **b** authors' method, **c** the algorithm described in [4]

comparison, the authors' algorithm gives better results in the initial stages of eye diseases than the algorithm presented in [4]. Besides, the solution proposed in [4] cannot detect small changes and lesions nearby large exudates, whereas authors' method can detect pathological changes in both of these situations.

Authors' database is continuously expanding. In the near future, the authors are planning to take into consideration lesions nearby optical disk and improve the accuracy of small exudates detection.

Acknowledgements This work was supported by grant S/WI/1/2013 from Bialystok University of Technology and funded with resources for research by the Ministry of Science and Higher Education in Poland.

References

1. Eadgahi, M.G.F., Pourreza, H.: Localization of hard exudates in retinal fundus image by mathematical morphology operations. J. Theor. Phys. Cryptogr. **1** (2012)
2. Jaafar, H.F., Nandi, A.K., Al-Nuaimy, W.: Automated detection and grading of hard exudates from retinal fundus images. In: 19th European Signal Processing Conference—EUSIPCO, pp. 66–70. (2011)
3. Jaya Kumari, C., Maruthi, R.: Detection of hard exudates in color fundus images of the human retina. In: International Conference on Communication Technology and System Design, Procedia Engineering, pp. 297–302. (2011)
4. Rokade, P., Manza, R.: Automatic detection of hard exudates in retinal images using haar wavelet transform. Int. J. Appl. Innov. Eng. Manag. **4**(5) (2015)
5. Kekre, H., Sarode, T., Parkar, T.: Hybrid approach for detection of hard exudates. Int. J. Adv. Comput. Sci. Appl. **4** (2013)
6. Stankiewicz, A., Marciniak, T., Dąbrowski, A., Stopa, M., Rakowicz, P., Marciniak, E.: Denoising methods for improving automatic segmentation in OCT images of human eye. Bull. Pol. Acad. Sci. Tech. Sci. **1** (2017)
7. Turchin, I., Sergeeva, E., Dolin, L., Kamensky, V., Shakhova, N., Richards-Kortum, R.: Novel algorithm of processing optical coherence tomography images for differentiation of biological tissue pathologies. J. Biomed. Opt. **10**(6) (2005)
8. Somfai, G., Tatrai, E., Laurik, L., Varga, B., Olvedy, V., Smiddy, W., Tchitnga, R., Somogyi, A., DeBuc, D.: Fractal-based analysis of optical coherence tomography data to quantify retinal tissue damage. BMC Bioinform. **15**, 295 (2014)
9. Świebocka-Więk, J.: The detections of the retina's lesions in Optical Coherence Tomography (OCT). In: CITCEP—Proceedings of the Congress on Information Technology, Computational and Experimental Physics, Kraków, Poland, 18–20 December. (2015)
10. Srinivasan, P., Kim, L., Mettu, P.S., Cousins, S.W., Comer, G.M., Izatt, J.A., Farsiu, S.: Fully automated detection of diabetic macular edema and dry age-related macular degeneration from optical coherence tomography images. Biomed. Opt. Express. **5** (2014)
11. Chaki, N., Hossain Shaikh, S., Saeed, K.: Studies in computational intelligence In: Exploring Image Binarization Techniques, vol. 560. Springer India, (2014). ISBN: 978-81322-1906-4

Typing Signature Classification Model for User Identity Verification

Tapalina Bhattasali, Rituparna Chaki, Khalid Saeed and Nabendu Chaki

Abstract Typing pattern is a behavioral trait of user that is simple, less costly, and workable at any place having only computing device. In this paper, n-graph typing signature is built during user profiling based on keyboard usage pattern. The main aim of this paper is to increase inclusion of number of typing features (both temporal and global) during decision generation and to simplify the procedure of considering missing typing patterns (various monographs, digraphs, etc), which are not enrolled before. A modular classification model collection–storage–analysis (CSA) is designed to identify user. Typing signature becomes adaptive in nature through learning from environment. Module 1 is used for pattern acquisition and processing, and module 2 is used for storage, whereas module 3 is used for analysis. Final decision is generated on the basis of evaluated match score and enrolled global parameters. Proposed CSA model is capable to reduce space and time overhead in terms of dynamic pattern acquisition and storage without using any approximation method. A customized editor HCI is designed for physical key-based devices to build our own data set. Proposed CSA model can classify typing signature of valid and invalid user without incurring high overhead.

Keywords Classification model · Typing signature · n-graph
Wildcard character · Identity verification

T. Bhattasali (✉) · R. Chaki · N. Chaki
University of Calcutta, Kolkata, India
e-mail: tapalina@ieee.org

R. Chaki
e-mail: rchaki@ieee.org

N. Chaki
e-mail: nabendu@ieee.org

K. Saeed
Bialystok University of Technology, Bialystok, Poland
e-mail: k.saeed@pb.edu.pl

© Springer Nature Singapore Pte Ltd. 2018
R. Chaki et al. (eds.), *Advanced Computing and Systems for Security*,
Advances in Intelligent Systems and Computing 666,
https://doi.org/10.1007/978-981-10-8180-4_4

1 Introduction

In order to control resource access effectively, reliable identity verification mechanism needs to be used. User identity based on biometric features is more efficient than use of only password. Biometric techniques provide a solution to ensure that the desired services are accessed only by a legitimate user and no one else. The main reason behind high reliability of biometric features to represent user's identity compared to many other traditional mechanisms is that it cannot be stolen like password. Trustworthiness of low cost password-based authentication can be increased by analyzing typing pattern of users. It is a behavioral nature which can be captured by the way individual types on a keyboard.

Several studies [1, 2] show that individual's typing patterns are stored as template and can be effectively used for identification. Timing vectors are mainly used to classify patterns as valid or invalid.

Various features extracted from typing events can be divided into temporal features and global features. Temporal features represent timing data for typing of specific key. These features are calculated based on the time stamps when the key is pressed and released. Global features refer to typing pattern of the user, such as frequency of errors, frequency of using control keys such as caps lock, num lock, shift, alt (e.g., left or right), overall typing speed. It is cheaper to implement than other biometrics as no additional pattern acquisition device is required. Typing pattern can be classified as fixed [3] and free [4]. Fixed pattern is based on pre-defined content, which makes it more prone to forgeries. On the other hand, free pattern [5] is based on random typing on keyboard, which makes it more challenging for user profile creation.

This work is extended version of one of our previous works [6] on modular classification model to validate typing signature pattern [7]. This paper considers both fixed and free and long and short patterns. The major contribution of this work is to use the concept of wildcard character ('*') for n-graph (where n = 1, 2, 3,..., m) creation during feature processing at module 1. Average key latencies are extracted from frequently used n-graphs (grouped on the basis of wildcard character) to deal with the n-graphs which are not available during enrollment. Use of generalized wildcard character-based n-graph for used patterns (patterns which are frequently used) enhances flexibility. Wildcard characters also reduce the sparse entries of template to improve performance of a classification model. Here, several global features (typing length, frequency, and count of using different types of patterns) are also considered during decision generation as outer layer of pattern matching along with temporal features (key time, typing duration, time per key pressed, typing speed for both fixed and free patterns) to enhance accuracy of classification model. Here, typing patterns are collected through HCI interface.

The remaining part of this paper is structured as follows: Section 2 presents a survey of some well-known research works on this domain. Section 3 presents research gap analysis, Sect. 4 introduces proposed classification model to validate typing signature for user identity verification, and Sect. 5 presents the analysis part. Finally, Sect. 6 concludes the paper.

2 Literature Survey

In general, keystroke analysis [8–10] is based on the traditional statistical analysis or pattern recognition techniques. Drawbacks of both neural networks and statistical methods in terms of search times are identified. It is claimed that performance of keystroke-based authentication [11] is better than vein pattern recognition and is similar to fingerprint and voice recognition for Internet-based authentication. Several classifiers are used with the trade-off between computation and performance. In this paper, only standard keyboard is considered for user data inputs as in the works [10, 12, 13]. There are several works on keystroke authentication based on either fixed text or free text. Users can be identified using either one-time verification or multi-time verification (continuous or periodic mode). In [14], key press interval is taken as a signature identifier. Implementation is platform-independent and does not require excessive computational power. However, this logic deals with fixed text, and data are statistically analyzed to determine keystroke patterns.

Ahmed et al. [5] proposed a free-text keystroke dynamics-based authentication, where raw data (flight time for digraph, dwell time for monograph) are collected, processed, and converted to digraph and monograph formats by using approximation method to consider missing patterns during enrollment. To illustrate the concept, we have briefly presented the procedure of digraph approximation for missing digraph.

Average flight times are computed for the digraphs ending with given characters.

Digraph key orders are computed by sorting average flight times.

Missing digraphs are estimated using digraph key order and considered as input of digraph neural network.

Outliers are removed from monograph and digraph sets during enrollment by Pierce's condition. After removing outlier, they are passed through sorting modules, which process data and calculate mapping tables for future use as a part of the signature. Missing digraphs need to be approximated by calculating average flight times for all the digraphs in the provided sample. User behavior for monographs is a 2-D relation between key order and its dwell time. Weights of the trained networks are also considered as a part of user's signature.

Sometimes, nature of variations between multiple valid keystroke entries contains sufficient discriminatory information to improve keystroke authentication [9]. Variation in typing sequences is independent of typing proficiency unlike other parameters. Variations in the event sequences decrease significantly, if users are familiar with typing of a specific string. Collected raw data including pressed key, time stamps of the key events, IP address, browser type, date and time of submission are submitted to the back-end server. Any key can be typed for 12-character-long password having format of "SUUDLLLLLDUUS", where S is any symbol, U stands for uppercase letters, L stands for lowercase letters, and D stands for numeric digit. As for example, user types string "HJealth", instead of "Health". One possible event sequence for this is {RSdHduRSu, LSd, J, LSu, BKdu, edu, adu, ldu, tdu, hdu}. Here, use of either right shift (RS) key or left shift (LS) key to type uppercase letters results into multiple events. Sliding window technique is considered on each subject pair to calculate the percentage of unique event sequences.

Users' typing patterns are also continuously monitored for authentication [10]. To reduce dimensionality, a feature vector is extracted from each input stream of keystrokes (session) and digraphs are clustered based on the temporal features. Although digraphs and their corresponding interval times are analyzed, authors claimed that it can work well with any n-graph and their temporal features as well as with any classification algorithm. This clustering logic shows better performance than k-means algorithm. Optimal number of n-graphs is required for cluster formation to show accuracy.

It is observed from the above study that larger sample size gives rise to better accuracy in case of free-text authentication [15]. However, shorter enrollment period is better suited for the security perspective. Major requirement of keystroke accuracy is to include greater number of participants and collect multiple samples for a long period of time. It is found that if users are unable to log in, or accepted at low rate, their keystroke patterns are inconsistent in nature, which may increase false detection.

3   Research Gap Analysis

According to the literature survey [5, 8–10], a few generalized issues are identified to draw more attention of the researchers.

Most of the keystroke-based authentication procedures consider only the temporal parameters like dwell time and flight time from the users' typing patterns, which may vary with time and are not capable to produce accurate result. There are still few works that also consider global parameters such as typing sequences, count of errors during typing, habit of typing, stylometry to increase reliability of users' typing profile. Accurate authentication is guaranteed through use of larger sample size, which is not always available. Run-time verification of random typing patterns still remains an open research issue.

Although there are several works which can be used as a base of future research, we have found Ahmed et al.'s [5] work more interesting. It is considered as a base of our research work from the perspective of dynamic pattern collection and storage as they have provided a solution on:

How to match typing features of monographs and digraphs of a user during verification, which were missing during enrollment?

A few issues addressed by Ahmed et al. are considered here as background of our research work. In Ahmed et al. [5] work:

- Limitations of enrollment process are removed by approximation technique known as sorted time mapping (STM). Only flight time is considered for digraph approximation, and dwell time is considered for monograph approximation. Missing digraphs are approximated by calculating average flight times, and monographs are considered as a 2-D relation between key order and its dwell time.
- Separate mapping tables need to be generated and stored, and separate neural network classifier models need to be designed for each type of mapping table.
- Samples are passed through sorting modules before generation of mapping tables.
- Outliers are removed by Peirce's criterion based on statistical analysis of the Gaussian distribution without depending on collected samples.

Therefore, it can be said that Ahmed et al. proposed a flexible and adaptive logic, which may give rise to computation-intensive procedures that need to be simplified.

3.1 Problem Statement

On the basis of background of our work, we can state our research problem as follows:

How to analyze random typing patterns without maintaining separate templates and separate classification models for different graphs (digraphs, monographs, etc) along with minimizing computational overhead in terms of space and time and reducing the probability of false detection of typing signature without using approximation method?

3.2 Scope of Work

Scope of our research work can be summarized as follows:

- To consider generalized n-graph signature instead of considering monograph signature and digraph signature separately.
- To include more typing features (both temporal and global) of users.
- To reduce computational complexity in terms of space and time overhead for pattern acquisition and storage.
- To increase reliability of outlier detection based on collected pattern.
- To include simple normalization procedure.
- To design efficient classifier to identify valid or invalid user based on typing signature patterns.

4 Proposed Model

Our objective is to design a modular classification model, which can classify valid and invalid typing signature of users based on random typing patterns [6, 16] to ensure flexibility, reduce requirements of storage space and processing time. This model uses typing pattern as input and produces a decision based on pattern matching. Typing signatures are created using temporal parameters of typing patterns and weight of network model used by classifier. Classification model is activated during enrollment phase and verification phase. Enrollment phase includes data acquisition, feature extraction, template generation, storage, and learning. Verification phase is further divided into pattern matching and decision making. Terminologies used in this paper are presented in Table 1.

4.1 Typing Signature Classification Model

Typing signature classification model CSA includes module 1 (C) for collecting data, module 2 (S) for storage, and module 3(A) for analysis. CSA model is designed based on both functional (F) and non-functional (NF) requirements. F has higher priority compared to NF. F includes identity verification, whereas NF includes anomaly detection (classification between human and bot to avoid synthetic forgeries [17]). Identity verification includes two subclasses: distinction class (1: m verification, where n = m) and authentication class (1: 2 verification, where n = 2). Outputs of

Table 1 Terminologies

Terminologies	Meaning
Match score generator	Considers classifier output, average deviation, threshold, biasness to generate match score
Decision maker	Checks match score and relevant temporal and global parameters to check if claimed user is an authentic user or not
P_{in}	Number of processing unit in input layer is set to number of selected features of timing vectors
P_{out}	Number of processing unit in output layer is set to 1, because the output is known, i.e., either valid (1) or invalid (0)
P_{hi}	Number of processing unit in hidden layer is set to (Pin + Pout)/2
α	learning_factor
η	adjust_factor
wt_{diff}	Difference in weight update
th_err	threshold_error
th_success	threshold_success
$node_{input}$	Number of input nodes

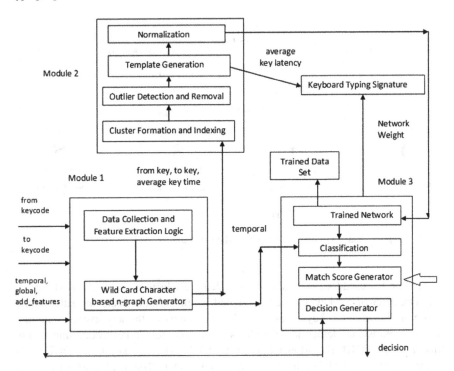

Fig. 1 Three modules of CSA model

classification model are categorized into two classes: valid class and invalid class. Valid class is used to check whether user is claimed one or not (authentication needs to check only profile of claimed user) and whether user is enrolled before or not (distinction needs to check all the profiles enrolled before). Invalid class may include user not possessing claimed identity (having no malicious intension), imposters (having malicious intension), bots (valid user's profile is generated to behave as valid user artificially). Figure 1 represents three modules of CSA model.

Typing signature classification model includes two tuples: {keyboard_typing_signature, add_feature}. Reference signature includes user-id, template vectors of frequently used typing patterns (used patterns), neural network weight. Typing patterns can be updated with time based on feedback path of classification model. Network weights are updated through learning phase. add_feature includes several temporal (which are not included for signature generation) and global features required during decision generation. Typing pattern classification is defined as a model having the following tuples: {fixed pattern, free pattern, long pattern, short pattern, temporal features, global features}. It is the example of multi-feature model. It is assumed that CSA model blocks any user for that time after two consecutive invalid attempts. Modules of our classification model are presented below.

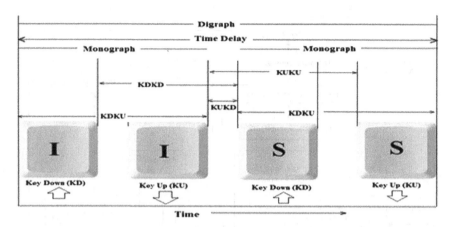

Fig. 2 Temporal data of keystroke for n-graph where n = 2

Module 1 is used for raw data collection, feature selection, and feature extraction. Any character of the keyboard can be typed. Application collects and submits the following information keys pressed and time stamps of the key events, date and time of submission, device-id, IP address. Raw data include key code, time stamp, time interval (KDKD (key-down, key-down), KUKU (key-up, key-up), KDKU (key-down, key-up), KUKD (key-up, key-down)). Time interval data of keystroke analysis (in milliseconds) are considered for n-graph. Figure 2 represents keystroke data for n-graph where n = 2.

- KDKU—Time gap between key press and key release (dwell time).
- KUKD—Time gap between release of previous key and press of next key (flight time).
- KDKD—Time gap between two successive key presses.
- KUKU—Time gap between two successive key releases.
- Key time—Time gap between key press of first key and key release of last key of n-graph.
- Average key time—Average of key times of n-graph.
- Average key latency—Average of all average key times of all same categories n-graph.

Typing patterns are classified into two categories: used patterns and rare patterns. Used pattern includes alphabet keys, whereas numeric and special characters are considered as rare patterns. Typing speed is calculated from the length of a string divided by key latency. Other parameters are count of pressing backspace, probability of using shift key or caps lock. It considers user's habit to type a sequence of string among all available possibilities; e.g., to type "Health", one typing sequence is

Table 2 Sample data during feature extraction

n-graphs (number of keys)	From key code	To key code	Key latency (ms)
and (3)	97	100	40
A (1)	15	15	20
is (2)	105	115	33

Table 3 Processed sample data

From key code	To key code		n-graph	Key time (ms)
97	100	a*d	and (3) around (1) acid (4)	• T1 • T2 • T3

Table 4 Reduced sample data

From key code	To key code	String (n-graph)	Average key time (ms)
97	100	a*d	(T1 + T2 + T3)/3

$<shift_{down}, h_{down}, h_{up}, shift_{up}, e_{down}, e_{up}, a_{down}, a_{up}, l_{down}, l_{up}, t_{down}, t_{up}, h_{down}, h_{up}>$. There are various other combinations to type it. User's sample feature vector is presented in Table 2.

In the proposed logic, concept of wildcard character is used. n-graph-based logic considers "from key" and "to key" of any string. As for example, average key latency of A-A (monograph) and average key latency of a-d (digraph, n-graph) are considered. n-graph uses concept of wildcard character, i.e., where string is "a*d", '*' is wildcard character, starting character is 'a', and ending character is 'd', and the key time (kt) is calculated by considering time difference between first key and last key of n-graph, divided by n number of strings included within "a*d". Key latency (kl) sums up all available average key time (akt) of a specific wildcard character-based n-graph. Average key latency (akl) is used to reduce the size of the sample, which is a major requirement to transmit data in client–server environment. The concept of extracted feature is given in Tables 3 and 4.

Data Collection and Feature Extraction Logic

Input: U number of users, S number of samples per user

Output: matrix for each user containing from key code, to key code, average key
latency in milliseconds

begin

for i=1 to U **for** j=1 to S

  capture_event (key_press, key_release)

  if $(n+1)^{th}$ key code = 32 then, // space as string separator

    ngraph = concat($[1,2,...,n]^{th}$ key)

      from_keycode = key code of 1st key

      to_keycode = key code of nth key

      write ngraph

      write 1st key $time_{keypress}$

      write n^{th} key $time_{keyrelease}$

  if 1st key $time_{keypress}$ < nth key $time_{keyrelease}$ then,

    kt= $|n^{th}$ key $time_{keyrelease}$ -1st key $time_{keypress}|$ *//key time*

  else

    exit

  if 1st key = 'x' and n^{th} key = 'x' then,

    $ngraph_{wildcard(x^*)}$ = {1^{st} key} $\cup$ {'*'}

    // such as "a"- start key 'a' and end key 'a'*

  if 1st key = 'x' and n^{th} key = 'y' then,

    *// such as "a*d"- start key 'a' and end key 'd''*

  $ngraph_{wildcard(x^*y)}$ = {1^{st} key} $\cup$ {'*'} $\cup$ {n^{th} key}

  akt=($|n^{th}$ key $time_{keyrelease}$ -1st key $time_{keypress}|$)/n *// average key time*

  endfor

  p =count ($ngraph_{wildcard(x^*)}$)

  q =count ($ngraph_{wildcard(x^*y)}$)

  akl for $ngraph_{wildcard(x^*)}$ = $(\sum akt)/p$

  akl for $ngraph_{wildcard(x^*y)}$= $(\sum akt)/q$

  creatematrix (l×l)

  add_item(matrix_id, from_keycode, to_keycode, akl)

  endfor

end

Module 2 is used for cluster formation from sample data, outlier detection (noisy data removal), normalization, template generation, and storage. This module is used for grouping filtered data to reduce sample size and to make searching faster. There are two major differences of proposed cluster formation with k-means algorithm. The first is that the number of clusters does not need to be specified in advance in this approach. The second is that indexing is used to reduce time complexity of k-means

[18] algorithm. Index database is used to search the location of a sample in a faster way. Clusterization of user's data is based on temporal information of used and rare patterns. Each cluster is partitioned into subclusters according to number of typable characters within a string. Intra-cluster difference (difference between a centroid and the sample data) and inter-cluster difference (difference between centroids of two clusters) are calculated to differentiate between valid and invalid data samples. Intra is used to measure the compactness of the clusters. Inter is used to measure the separation of the clusters. The logic of cluster formation is presented below.

Indexed Clustering Logic

Input: Dataset ds containing S data samples for U users
Output : p number of clusters and p number of index
begin
   pick randomly k of n data points pt as cluster centre(cc)
for cc = 1 to k,
for pt = 1 to n,
  calculate d = distance(x1,y1,x2,y2) // *distance between pt and cc*
  calculate d_{min}
    add (cc, pt, d_{min})
endfor
endfor
for each cc data point m_i and for each non-cc data point o_i,
 swap m_i and o_i
endfor
 config_cost= compute(pt, d, cc)
for each cluster c,
   closeness= compute (x_i, x_{mean},pt, n)
 calculate config_cost$_{min}$
   set cc = newcc(config_cost$_{min}$)
   calculate intra   // *Intra-Cluster distance*
   calculate inter   // *Inter-Cluster distance*
   if no change in cc then,
    continue
    generate index for each cluster
  validity= intra/inter
  endfor
end

Outliers of the extracted features are detected to remove noisy data during cluster formation at module 2 by distant outlier detection logic presented below.

Distant Outlier Detection Logic

Input: Dataset ds containing S data samples, p number of clusters

Output: Set of Outliers OUT

begin

    for each cluster C_i

    set $radius_i$ = radius of cluster C_i

    for each point o_j in $ds_i \in C_j$

    if d>= $radius_i$ then,

        OUT = OUT $\cup$ o_j

    else if $o_j$belongs to both C_i and C_j

        OUT = OUT $\cup$ o_j

    endfor

    endfor

    delete outlier set OUT

end

Outlier set includes any pt, which is not included within any cluster, which is included in more than one cluster, which is on the boundary of the cluster. Outlier removal technique proposed in [8] is simplified here by using standard deviation method to remove noisy data during cluster formation.

Template is generated from the clustered data. There are several clusters formed to record user's typing patterns. Template stores the keystroke signatures of a user. Temporal data of used patterns of keystroke (alphabets) are normally used for template generation. Single template for each user that enables proposed logic to distinguish between legitimate user and impostor with minimum rate of error is considered. A 52×52 matrix is formed to represent time delay to press a key or time delay between two keys. Both upper case and lower case characters are considered. Diagonal elements represent monographs, and non-diagonal elements represent digraphs. Sparse entries are discarded to reduce template size, and frequently used n-graphs (monographs and digraphs) are considered during data normalization. $kl_{i,j}$ represents here average key latency for monographs, digraphs as well as n-graphs. Sample template in matrix format is provided in Fig. 3.

Each cell of the template matrix contains key latency. In matrix, the common cell between A and M includes average of all time latency of n-graphs whose starting character is 'A' and ending character is 'M'. Template is generated in such a way so that size of the template is reduced. Data are normalized and changed into a form recognizable by module 3 classifier. Data normalization logic is presented below using MinMax [19] logic.

***Verification Logic of Sample by Trained N2BP* Input:**
Input: Timing Vector TV_i of claimed user U_i
Output: Classification Decision either valid (1) or invalid (0)
begin
    initialize sc to 0
    for each component in input vector TV_i of claimed User U_i
        calculate Signature $Sig_{(Ui)}=TV_i \times W_{i,j}$
                     // $W_{i,j}$ is weight of the network
        calculate abs_dif= $[(TV_i-W_{i,j})+ (RS(U_i)- Sig(U_i))]$
                    // RS(Ui) represents stored reference signature
    endfor
    calculate tolerance_limit=adjust_factor × variability
    if abs_dif<= tolerance_limit then,
        increment sc   *//success counter*
    if sc / node$_{input}$>= th_success and targetval=outputval
              or targetval-outputval<0.0001 then,
        validity = 1
        update weight
    else
        validity = 0
    end

N2BP model may not generate exactly 0 or 1 for using sigmoid function. If allowable tolerance level of N2BP model <= 0.1, it is treated as 0, and if it is >=to 0.9, it is equal to 1. If testing of data for N2BP model results within 0 and 1, then testing data is treated as valid (1); otherwise, it is treated as invalid (0). Match logic evaluates score only on the basis of classification accuracy, average key latency, average deviation, weight, and biasness of the network.

Average deviation [6, 16] for n-graph is calculated as follows:

$$\Delta_{ngraph} = \text{difference} (RSKL_i, kl_i, N, 100)$$

where N is the total number of data set, kl_i is the key latency for the ith sample in user's session, $RSKL_i$ is resulting from claimed user's n-graph neural network model. Average deviation represents similarity in user's behavior in a session to the valid user's behavior. Lower number represents high similarity to ensure that session belongs to the same user. Match score is calculated as follows:

$$\text{match\_score} = 1, \quad \text{if } m \text{ is} <= \text{th},$$

where $m = \Delta \times b$, and b is biasness, this threshold

$$= 0 \quad \text{otherwise}$$

Table 5 Fixed pattern: .tie5Roanl (10-graph)

uid	Session	From key code	To key code	n-graph	Typing gap (in second)	Character per second (cps)
u1	s1	48	108	10	(Absolute value of KD of '.'-KU of 'l')/10-3	Typing gap/number of characters

Decision logic finally takes the decision based on match score and relevant features including parameters such as rare data pattern, user nature of typing to verify claimed identity.

5   Performance Analysis

Performance of proposed work is analyzed in MATLAB R2012b. At present, analysis is done in laboratory environment [20–22]. QWERTY keyboard is considered for data collection through Windows API. All the participants use same HP laptop (2.4 GHz Pentium 4 processor, with 2 GB of RAM, and running Windows 7 (64 bits)).

In order to analyze the proposed method, data are collected locally through user-friendly interface. Typing events are collected from QWERTY keyboard through Windows API. Besides considering popularly used parameters like true acceptance rate (TAR), true rejection rate (TRR), false acceptance rate (FAR), false rejection rate (FRR), equal error rate (EER), and accuracy, we have considered here parameters like failure to capture (FTC), average false acceptance, average false rejection.

FTC = (erroneous capture / total number of typing stroke)

FTC rate (%) = FTC × 100

Average False Acceptance = ∑Number of Imposters accepted as Genuine user for a specific threshold / count of false acceptance for that threshold

Average False Rejection = ∑Number of Genuine user rejected for a specific threshold / count of false rejection for that threshold

5.1   Analysis of the Collected Data

Fixed texts are treated as a password. Two passwords can be used alternatively: (i) .tie5Roanl (10-graph) and (ii) try4-mbs (8-graph). Another fixed pattern is used as optional pass code: —a quick brown fox jumps over the lazy dog: —containing all 26 alphabets (used pattern). Free texts are collected from users as they type sample text within 100–150 words. It can vary from user to user. Common texts are considered to differentiate between two users (inter-variance), whereas uncommon texts are considered to analyze typing pattern of the same user (intra-variance) (Tables 5 and 6).

Table 6 Fixed pattern: try4-mbs (8-graph)

uid	Session	From key code	To key code	n-graph	Typing gap (in second)	Character per second (cps)
u1	s2	116	115	8	Absolute value of KD of 't' -KU of 's')/10^{-3}	typing gap/number of characters

Table 7 Temporal data of free text of uid AB for session starts at 1:24:31 AM

Key	Key code	Key event time (ms)
I	73	22746326.54
space	32	22746326.70
a	97	22746326.94
m	109	22746327.16
space	32	22746327.31
a	97	22746327.45
space	32	22746327.63
d	100	22746327.78
o	111	22746327.92
c	99	22746330.23
t	116	22746330.55
o	111	22746330.85
r	114	22746331.14
	46	22746331.42

Table 8 Data of different users for fixed text and free text

uid	Fixed text typing duration	Time per key pressed for fixed text	Length of free text	Free-text typing dura-tion	Time per key pressed for free text	Free-text typing speed
AB	14	1.4	69	93	1.34782608695652	0.741935483870968
PB	8	1	67	74	1.1044776119403	0.905405405405405
AD	5	0.625	85	69	0.811764705882353	1.23188405797101
MP	14	1.75	71	146	2.05633802816901	0.486301369863014
MB	16	1.6	68	192	2.82352941176471	0.354166666666667

Tables 7, 8, and 9 represent raw data collected from different users at module 1.

In typing signature analysis, there is a probability of failures during data acquisition. These failures are due to (i) users who want to type faster than their ability, (ii) users who are disturbed by the environment, (iii) users who know that their typing times are being saved, (iv) users who may not be familiar with using the keyboard. Failure to capture (FTC) can be calculated from erroneous captures and a total num-

Table 9 Typing patterns of different users (global parameters)

uid	Used patterns	Alphabets	Shift	Space	Rare patterns
AB	53	53	4	13	3
PB	154	144	12	34	12
AD	147	139	13	33	16
MP	49	49	0	12	2
MB	199	181	19	45	20

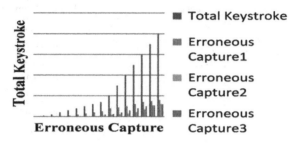

Fig. 4 Total keystroke versus erroneous capture of keystroke

Table 10 Free-text sample entry

n-graph	Common string	From key code	To key code	Average key latency (ms)	
				User 1	User 2
3	AND	65	100	12.5	20.1

ber of keystrokes. FTC in the proposed work is low, as there is less requirement of typing fixed text. It reduces the probability of mistakes. User can type text freely and can use any key of the keyboard during entire session. Figure 4 represents plot of total keystroke and erroneous capture. FTC for user 1 (erroneous capture1) is found to be 0.2, whereas FTC for user 2 (erroneous capture2) and user 3 (erroneous capture3) is 0.155 and 0.15, respectively. FTC of user 1 is analyzed for general fixed text-based authentication, whereas FTC of user 2 and user 3 is analyzed according to the proposed work. It shows that FTC rate is around 20% in general, whereas it is around 15% for our proposed work. It proves that FTC rate is improved in proposed work.

Sample size includes 35 users. Verification is done 10 times at each threshold level. Table 10 represents free-text sample entry.

In Fig. 5, average key latency is plotted against five different users for typing the same strings. Timing data of typing same n-graph vary for five different users.

Average false acceptance and average false rejection are calculated from Tables 11 and 12.

In Fig. 6, trade-off between average false acceptance and average false rejection is plotted as average false detection against different threshold levels. It shows that

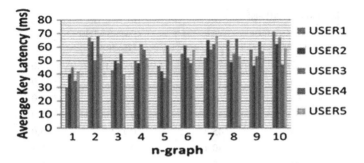

Fig. 5 Plot of average key latency for same n-graphs

Table 11 Average false acceptance and false rejection

Threshold	Number of invalid users accepted out of total sample size at different sessions	Number of valid users rejected out of total sample size at different sessions
5	<12,11,12,11,11,11,11,12,11,12>	<0,0,0,0,0,0,0,0,0,0>
10	<10,11,10,9,11,11,10,11,11,11>	<0,0,0,1,0,0,1,0,0,0>
15	<9,10,9,8,9,9,8,10,9,9>	<0,0,2,1,0,2,0,1,0,0,>
20	<7,8,7,8,9,7,7,8,7,7>	<0,0,2,1,2,1,0,0,2,0>
25	<7,7,6,7,6,7,7,6,7,8>	<1,3,2,2,0,0,0,0,0,0>
30	<6,6,7,6,6,6,6,6,6,6>	<3,0,0,2,3,0,0,0,0,0>
35	<5,5,6,7,5,6,4,5,5,5>	<3,0,0,0,0,3,0,3,0,0>
40	<3,4,3,4,4,5,4,3,4,4>	<1,3,2,1,0,0,0,2,0,0>
45	<2,3,3,4,2,3,3,3,4,3>	<2,1,0,0,1,0,2,3,0,0>
50	<2,3,2,1,3,2,3,2,3,1>	<2,0,0,2,0,4,0,0,1,0>
55	<1,2,1,1,2,3,0,1,2,2>	<4,0,4,0,0,0,0,1,0,0>
60	<2,1,1,0,0,2,2,1,2,2>	<3,2,1,0,0,0,3,0,0,0>
65	<1,0,1,1,0,1,1,2,0,1>	<4,0,2,1,0,2,0,0,0,2>
70	<1,0,0,0,1,0,1,1,0,1>	<1,3,0,0,4,0,0,2,1,0>
75	<1,0,0,1,1,0,0,0,0,0>	<3,0,4,1,0,0,0,0,2,2>
80	<0,0,0,1,0,0,0,1,0,0>	<4,2,0,3,0,0,0,3,2,1>
85	<0,0,0,0,0,0,0,0,0,0>	<5,4,6,5,4,4,5,4,4,4>
90	<0,0,0,0,0,0,0,0,0,0>	<8,9,7,7,8,7,7,7,7,8>
95	<0,0,0,0,0,0,0,0,0,0>	<12,12,11,10,12,11,11,12,11,11>

FAR is high and FRR is low when the threshold level is low. However, reverse is true when the threshold is high (Table 13).

Figure 7 represents comparative analysis of FAR and FRR of the proposed work at different threshold levels along with FAR and FRR of Ahmed et al. work with digraph approximation and without approximation method. According to analysis result, it is seen that proposed model works far better than the work without using any approximation method and is almost equivalent to approximation method with

Table 12 Calculation of FAR and FRR

Threshold (%)	Average false acceptance	FAR (%)	Average false rejection	FRR (%)
5	11.26	75.1	0	0.0
10	10.5	70.0	0.15	1.0
15	9.0	60.0	0.6	4.0
20	7.5	50.0	0.75	5.0
25	6.78	45.2	0.752	5.0
30	6.06	40.4	0.768	5.1
35	5.28	35.2	0.918	6.1
40	3.78	25.2	0.927	6.2
45	3.01	20.1	0.93	6.2
50	2.25	15.0	0.966	6.4
55	1.51	10.1	0.976	6.5
60	1.26	8.4	0.982	6.5
65	0.768	5.1	1.07	7.1
70	0.468	3.1	1.08	7.2
75	0.33	2.2	1.22	8.1
80	0.163	1.1	1.5	10.0
85	0	0.0	4.5	30.0
90	0	0.0	7.5	50.0
95	0	0.0	11.26	75.1

Fig. 6 Average false acceptance and average false rejection trade-off with threshold level

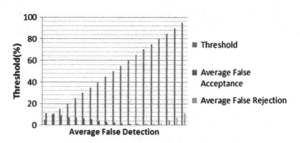

less overhead. Sometimes, proposed model even works better than the work using approximation method.

5.2 Discussion

The proposed work is based on generalized concept of n-graph-based classification model. Our classification enhances the flexibility compared to existing works [5, 8–10]. There is no requirement of maintaining mapping tables for monographs and digraphs or modeling the neural network separately [5]. It reduces time and space

Table 13 Comparative analysis of FAR and FRR

Threshold (%)	Ahmed et al. (with digraph approximation)		Ahmed et al. (without digraph approximation)		Proposed work (without digraph approximation)	
	FAR (%)	FRR (%)	FAR (%)	FRR (%)	FAR(%) (approx.)	FRR(%) (approx.)
5	71.671	0	98	0	75.1	0.0
10	68.443	0	96	0	70.0	1.0
15	56.184	0	84	5	60.0	4.0
20	52.201	0	80	6	50.0	5.0
25	48.291	0	78	8	45.2	5.0
30	44.093	0	65	10	40.4	5.1
35	38.149	0	60	10	35.2	6.1
40	29.021	0	45	10	25.2	6.2
45	19.428	0	35	15	20.1	6.2
50	16.09	0	25	15	15.0	6.4
55	12.162	0	15	30	10.1	6.5
60	7.308	0.082	10	40	8.4	6.5
65	3.851	1.711	9	50	5.1	7.1
70	0.92	1.711	8	50	3.1	7.2
75	0.92	4.82	5	60	2.2	8.1
80	0.0152	6.1	5	60	1.1	10.0
85	0.0152	12.82	3	70	0	30.0
90	0	48.6	2	80	0	50.0
95	0	91	2	90	0	75.1

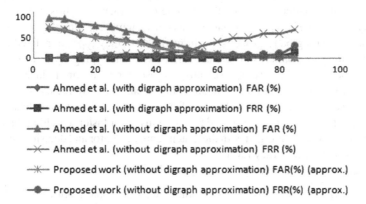

Fig. 7 Comparative analysis of FAR and FRR at different threshold levels

overhead. As the proposed work is based on the adaptive model of user's nature, it updates dynamically according to iterative learning process and enhances the flexibility. Feature extraction calculates average latency of n-graph in a very efficient way that reduces the feature set size. Simple clustering technique is used to group similar data samples in a cluster, where indexing technique is used for each cluster to reduce search time without enhancing the cost. Sparse entries in template are discarded to avoid operational complexity of classifiers. It may be discarded when it has no use; again, it can be added, according to the requirement, which consumes less resource. Odd data are removed in the simplest way compared to the other existing methods [5]. Data are normalized using a very simple logic. A reference typing signature of the user in the proposed model is based on temporal feature vector of user and the weight of network model, which can efficiently differentiate between two users. In summary, analysis of proposed model reveals that

- No mapping tables or separate neural network models for monographs and digraphs.
- Simple clustering technique groups similar data samples, where indexing technique is used for each cluster to reduce search time.
- Odd data are discarded based on intra-cluster and inter-cluster distance.
- Data are normalized using simple min-max logic.
- Both temporal and global features are considered for decision making.

6 Conclusions

The primary goal of the proposed model is to classify users accurately in a less complex manner. This does not introduce any additional costs and needs no additional hardware except the keyboard that every computer is equipped with. Dynamic pattern analysis is still a challenging issue. The proposed work focuses on this direction. Here, user's typing nature is captured besides considering temporal data of typing patterns. In module 1, typing patterns of users are recorded as they type short fixed patterns as well as long free patterns. Thus, the variations in the typing styles of individual users can be obtained easily. Sparse entries of user templates are reduced by using the wildcard character, and user's adaptability is enhanced by feedback path. This reduces the complexity faced by the classifiers. In module 2, clusters of data samples are formed and indexing is considered for each cluster for faster searching. In module 3, classifier decision is merged with the match score generator and decision generator, which enhance the accuracy level.

At a glance, the features of proposed model are as follows:

- Classification model considers temporal as well as global features of both static and dynamic patterns.
- Wildcard character ('*') is used to reduce the problem of matching patterns, which are not collected before.
- Size of template is reduced, and sparse entries are discarded to avoid processing overhead of classification model.
- Computation overhead is reduced along with maintaining accuracy for a limited set of samples in laboratory environment.

The authors believe that the proposed work could give better performance even in remote applications such as remote health care. The work is on to study various challenges in this domain and adjust the solution accordingly, so that it can be considered for further detailed processing and analysis.

Appendix

HCI User Interface
HCI Editor for Static Pattern Collection

HCI Editor for Dynamic Pattern Collection

References

1. Wang, Y., Du, G-Y., Sun, F.-X.: A model for user authentication based on manner of keystroke and principal component analysis. In: Proceedings of International Conference on Machine Learning and Cybernetics, pp. 2788–2792 (2006)
2. Balagani, K.S., Phoha, V.V., Ray, A., Phoha, S.: On the discriminability of keystroke feature vectors used in fixed text keystroke authentication 32(7), 1070–1080 (2011)
3. Gunetti, D., Picardi, C., Karnan, M., Akila, M., Krishnaraj, N.: Biometric personal authentication using keystroke dynamics: a review. J. Appl. Soft Comput. 11(2), 1515–1573 (2011). (Elsevier)
4. Gunetti, D., Picardi, C.: Keystroke analysis of free text. ACM Trans. Inf. Syst. Secur. 8(3), 312–347 (2005)
5. Ahmed, A., Traore, I.: Biometric recognition based on free-text keystroke dynamics. IEEE Trans. Cybern. 44(4), 458–472 (2014)
6. Bhattasali, T., Panasiuk, P., Saeed, K., Chaki, N., Chaki, R.: Modular logic of authentication using dynamic keystroke pattern analysis. ICNAAM, AIP Publ. Am. Inst. Phys. 1738, 180012 (2016)
7. Bhattasali, T., Saeed, K.: Two factor remote authentication in healthcare. In Proceedings of IEEE International Conference on Advances in Computing, Comunications and Informatics, pp. 380–381 (2014)
8. Giroux, R.S., Wachowiak, M.P.: Keystroke based authentication by key press intervals as a complementary behavioral biometric systems. In: Proceedings of IEEE International Conference on Man and Cybernetics, pp. 80–85 (2009)
9. Syed, Z., Banerjee, S., Leveraging, B.C.: Variations in event sequences in keystroke dynamics authentication systems. In: Proceedings of IEEE International Symposium on High-Assurance Systems Engineering, pp. 9–11 (2014)
10. Shimshon, T., Moskovitch, R., Rokach, L., Elovici, Y.: Clustering Di-graphs for continuously verifying users according to their typing patterns. In: Proceedings of IEEE Convention of Electrical and Electronics Engineers in Israel, pp. 445–449 (2010)
11. Karnan, M., Krishnaraj, N.: Bio password—keystroke dynamic approach to secure mobile devices. In: Proceedings of IEEE International Conference on Computational Intelligence and Computing Research, pp. 1–4 (2010)

12. Hu, J., Gingrich, D., Sentosa, A.: A K-Nearest neighbor approach for user authentication through biometric keystroke dynamics. In: Proceedings of IEEE International Conference on Communications, pp. 1551–1510 (2008)
13. Killourhy, K.S., Maxion, R.A.: Comparing anomaly-detection algorithms for keystroke dynamics. In: Proceedings of IEEE/IFIP International Conference Dependable Systems and Networks, pp. 125–134 (2009)
14. Araujo, L.C.F., Sucupira, L.H.R., Lizarraga, M.G., Ling, L.L., YabuUti, J.B.T.: User authentication through typing biometrics features. IEEE Trans. Signal Process. **53**(2), 851–855 (2005)
15. Killourhy, K.S., Kevin, S., Maxion, R.A., Roy, A.: Free versus transcribed text for keystroke dynamics evaluations. In: Proceedings of Workshop: Learning from Authoritative Security Experiment Results, pp. 1–8 (2012)
16. Bhattasali, T., Saeed, K., Chaki, N., Chaki, R.: Bio-authentication for layered remote health monitor framework. J. Med. Inf. Technol. **23**(2014), 131–140 (2014)
17. Jain, K., Ross, A., Pankanti, S.: Biometrics: a tool for information security. IEEE Trans. Inf. Forensics Secur. **1**(2), 125–143 (2001)
18. Kao, B., Lee, S.D., Lee, P.K.F., Cheung, D.W., Ho, W.S.: Clustering uncertain data using voronoi diagrams and r-tree index. IEEE Trans. Knowl. Data Eng. **22**(9), 1219–1233 (2010)
19. Xie, Q.Y., Cheng, Y.: K-Centers min-max clustering algorithm over heterogeneous wireless sensor networks. In: Proceedings of IEEE Wireless Telecommunications Symposium, pp. 1–6 (2013)
20. Giot, R., El-Abed, M., Hemery, B., Rosenberger, C.: Unconstrained keystroke dynamics authentication with shared secret. Elsevier Comput. Secur. **30**(1–7), 427–445 (2011)
21. Upmanyu, M., Namboodiri, A.M., Srinathan, K., Jawahar, C.V.: Blind authentication: a secure crypto-biometric verification protocol. IEEE Trans. Inf. Forensics Secur. **5**(2), 255–218 (2010)
22. Montalvao, J., Freirem, E.O., Bezerra, M.A., Garcia, R.: Empirical keystroke analysis in passwords. In: Proceedings of ISSNIP/IEEE Biosignals and Biorobotics Conference: Bio signals and Robotics for Better and Safer Living (BRC), pp. 1–6 (2014)

Part II
Image Processing

A Novel Technique for Contrast Enhancement of Chest X-Ray Images Based on Bio-Inspired Meta-Heuristics

Jhilam Mukherjee, Bishwadeep Sikdar, Amlan Chakrabarti,
Madhuchanda Kar and Sayan Das

Abstract Chest radiography is considered as one of the most important radiological tools in pulmonary disease diagnosis. Due to the generation of low contrast images of X-ray machines, the detection of the lesions is a difficult issue and prone to error for a radiologist. Hence, a contrast enhancement algorithm is an obvious choice to enhance the contrast of the image, thus increasing the accuracy of detection of the lesions. This paper not only proposes a new algorithm for contrast enhancement of digital chest X-ray images using particle swarm optimization (PSO), but it also introduces a benchmark dataset of digital chest radiographs to justify the supremacy of our proposed algorithm over that of state-of-the-art contrast enhancement algorithms.

Keywords Chest radiography · Contrast enhancement · PSO · Lung lesion
Benchmarking database

J. Mukherjee (✉) · A. Chakrabarti
A.K. Choudhury School of Information Technology,
University of Calcutta, Kolkata, India
e-mail: jhilam.mukherjee20@gmail.com

A. Chakrabarti
e-mail: acakcs@caluniv.ac.in

B. Sikdar
Institute of Engineering Management, Kolkata, India
e-mail: rahulsikder223@gmail.com

M. Kar · S. Das
Peerless Hospitex Hospital, Kolkata, India
e-mail: madhuchandakar@yahoo.com

S. Das
e-mail: sadnayas@gmail.com

© Springer Nature Singapore Pte Ltd. 2018
R. Chaki et al. (eds.), *Advanced Computing and Systems for Security*,
Advances in Intelligent Systems and Computing 666,
https://doi.org/10.1007/978-981-10-8180-4_5

1   Introduction

Lung cancer has become a serious health issue in recent days. Nearly 91% affected Indians are suffering death due to this disease [1]. The survival rate is increased if it is detected at an early stage. With the advancement of medical imaging technology, computer-aided diagnostic systems take a major role to predict this disease at an early stage. Although computed tomography (CT) images efficiently detect this disease at an early stage, this technology is not available throughout this country. Chest radiographs may become an alternative solution for this disease. Lung cancer is generally initiated by pulmonary nodules having a small white spot, visible in the lung parenchyma. Often these lesions are hindered by chest ribs. Besides, most of the X-ray scanners generate images that are of low contrast due to the presence of water in the human body, which makes interpretation of lesions a tedious job not only for a radiologist but also for a computer-aided diagnosis system. Although the increase in X-ray tube current can enhance the quality of the radiographs, it can also generate adverse effects on the human body. Hence, an appropriate contrast enhancement algorithm is required to enhance the image at a desired level.

In this paper, we have introduced a new chest X-ray dataset "Swash" for detection and prediction of malignancy in pulmonary nodules. Besides the dataset, we have also proposed a new methodology based on PSO for contrast enhancement of digital chest radiographs.

The rest of the paper is organized as follows: Sect. 2 discusses on the literature review in this domain. Algorithm and functions responsible for enhancement are illustrated in Sect. 4. Section 5 exhibits the result of this study. Finally, conclusions are drawn in Sect. 6.

2   Related Work

A digital medical image database was first introduced in digital mammograms known as digital database for screening mammography, which contains 2620 four-view screening digital mammograms [2]. A chest X-ray database of 247 scans of solitary pulmonary nodules has been proposed by Japanese Society of Radiological Technology (JSRT) [3]. Contrast enhancement increases the dynamic range of the image. There are many algorithms that can increase the contrast of an image to some extent. Histogram equalization (HE) [4] and linear contrast stretching (LCS) [5] are widely acceptable as they are very simple method to implement; however, these two are global contrast enhancement methods which have a tendency to incorporate noise in it. According to Chen and Ramli [6], HE increases the contrast using the middle value of the gray scale but not using the mean intensity value. This leads to a brightness preservation problem of histogram equalization. The above-mentioned problems can be overcome by adaptive histogram equalization (AHE) technique where the intensity value is increased based on the local properties of the histogram [7]. In another modified algorithm of HE, fuzzy histogram is used to overcome the inconsistency present in the gray level of brightness preserving dynamic histogram preceded by a smoothing procedure using Gaussian kernel, and finally, the gray level value is increased dynamically. Based on clipping histogram equalization technique, a novel contrast enhancement method is proposed in [8], where a clipping histogram is performed by mean and the median value of the histogram intensity adaptively. Using a region-based image enhancement technique, the contrast of the radiographs has been [9] increased by selecting the seed point and then intensity is enhanced adaptively. Another novel contrast enhancement methodology has been sketched out in [10], where benchmark images and low contrast digital mammograms are enhanced using shearlet domain. Computing the local statistical value of the input images, adaptive histogram enhancement technique is proposed in [11] to increase the quality of radiographs. Based on the local bihistogram equalization technique, an adaptive contrast enhancement methodology has been sketched out [12] to enhance the contrast of the MRI images. Using genetic algorithm (GA), a novel contrast enhancement methodology has been proposed in [13] to enhance the contrast of natural source images as it has more dynamic range, and the authors claim that GA can increase it to some extent. Using particle swarm optimization (PSO) technique, a novel contrast enhancement methodology has been sketched out in [14].

3 Data Description

Our target dataset images are described in Sect. 3.1. Presently, 120 chest X-ray of Swash dataset are available on (http://www.cu-coe.in/samples/X-ray/Data.zip). All the images are collected from Peerless Hospital from March 2015 to November 2016 for those patients who come for their routine check for lung cancer diagnosis. While collecting the patient data, we have taken those data that are confirmed through biopsy or cytology test. The patient edge lies between 30–75 in both male and female patients. Among 120 patients, 77 patients are male and rest are female. In conventional X-ray imaging procedure, the X-ray beam passes through the human body to project the shadow of hard tissue to project as shadow on the film. This film is then processed and printed to get the picture of the organ. In this study, we have used CR type of digital X-ray images where a sensor is placed behind the patient instead of placing X-ray films in conventional radiography. This sensor is next attached to the computer to create the digital radiography.

3.1 Sample Size Calculation

The size of the sample can be calculated as

$$n = \frac{z^2 * se(1 - se)}{d^2 p} \tag{1}$$

At 95% confidence level, $z = 1.96$, Se is the sensitivity; here, we considered it as 80%. d is the precession 20%, and p is the prevalence considered as 10%. Hence, the sample size of the dataset is 307 (Figs. 1 and 2).

3.2 Description of the Chest X-Ray Images

- Size: A pulmonary nodule with size < 3 mm is clearly benign. However, all the pulmonary nodules larger than this specified size do not have same chances of malignancy.

  - A lung nodule with size ≤ 4 mm has 0% chance of malignancy.
  - A lung nodule with size ≤ 7 mm has 1% chance of malignancy.

Fig. 1 Chest radiographs with nodule

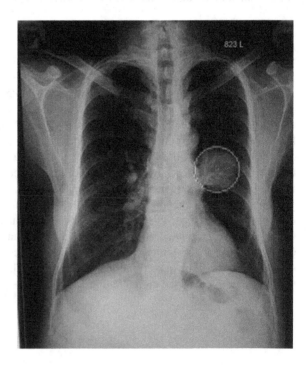

Fig. 2 Chest radiographs without nodule

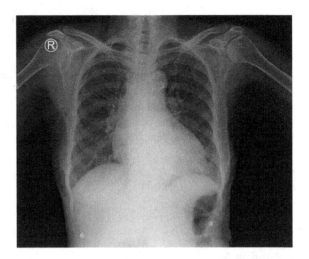

– A lung nodule with size ≤ 1 cm has 15% chance of malignancy.
– A lung nodule with size ≤ 2 cm has 40% chance of malignancy.

- Shape: The shape of pulmonary nodule is categorized into six groups, namely round, oval, lobulated, speculated, irregular, and ragged. The pulmonary nodule with shape round, oval are benign and speculated, irregular and ragged are clearly malignant where as pulmonary nodules with shape lobulated is belongs to any group depending upon another features.
- Margin: The outline pulmonary nodule is categorized into four groups, namely smooth, lobulation, speculation, and irregular. The first two indicate benignity, and rest two are malignant.
- Presence and pattern of calcification: Calcification is a process in which calcium is deposited in living cell. If the nature of calcification of a nodule is either central or popcorn or laminated, it is a sign of benign lung nodule, whereas if its nature is speckled or eccentric, it is malignant one.
- State of Abnormality

    – Truly benign: The histopathology report clearly shows that it has no chances of cancer.
    – Probably benign: According to radiological features, the pulmonary nodules have very few chances of malignancy.
    – Probably malignant: According to radiological features, the pulmonary nodules have chances of malignancy.
    – Malignant: All the reports clearly reveal that it has cancer.
        Adenocarcinoma: This type of lung cancer generates due to abnormal growth of adeno cell of human body.
        Squamous cell carcinoma: This cancer arises due to abnormal growth of squamous cell.

- Metastatic lung cancer: This type of cancer is generally initiated from other cancerous organs.

- Gender
- Position of the nodule: Human lung consists of five lobes. This confirm the localization of nodule on human lung.
- Coordinate position of the nodule:
- Types of Nodules

  Obvious
  Relatively Obvious
  Subtle
  Very Subtle
  Extremely Subtle

- Clinical Features

  Patient Age
  History of Malignancy
  Smoking History
  Hemoptysis

3.3   Annotation Process

In order to annotate the region of interest, we have designed a graphical user interface tool in MATLAB. Each radiologist marked on each subjects according to their expertise and visual interpretation considering a standard protocol of marking using this GUI tool through computer interface. All the radiologists mark on region of interest with a 5-mm interval on the boundary. This reduces inconsistency in marking. All the marks performed by each radiologist are stored in XML file along with slice number and its size. Each nodule candidate is assigned a unique identity number. They are assigned the number in increasing order from top left position to right bottom position. Each identification number remains the same in each slice in which it belongs. This procedure yields a single separate XML file for each subject.

3.4   Database Access Procedure

The original anonymized patient images in DICOM format along with its annotated image files of all subjects along with associated XML files and radiologists marked images of each patient have been uploaded to the following link (http://www.cu-coe.in/samples/X-ray/Data.zip) to download publicly and freely from the same link.

A registration procedure must be performed by each of the users before the sign-in process using users institutional email id. A unique password is automatically resent to the user's email id after verifying his/her affiliations and requirements. This password is valid for 15 days. The user is allowed to download a sample dataset without registration to check if it fulfills the, before going for a complete download after registration.

4 Methodology

In this section, we have described the way PSO is used to enhance the contrast enhancement of an image. Figure 3 illustrates the flow of the study.

4.1 Transformation Function

A low contrast image reveals that it has low intensity value. A transformation function can be applied on a low contrast image to enhance the gray value of the image.

Let us consider a grayscale image with size $M \times N$ whose intensity value can be enhanced by a transformation function T based on both global and local enhancement methodology. The enhanced image matrix can be defined as:

$$x(i, j) = T[y(i, j)] \tag{2}$$

where $y(i, j)$ is the intensity value of the input image at position (i, j) and $x(i, j)$ is the intensity value of enhanced image at same position [14].

Local transformation function is applied on a user-defined window size $n \times n$ which can be defined as

$$x(i, j) = S(i, j)[y(i, j) - c \times p(i, j)] + p(i, j)^a \tag{3}$$

where n is the local mean and defined as

$$p(i, j) = \frac{1}{n \times n} \sum_{i=0}^{n-1} \sum_{j=0}^{n-1} y(i, j) \tag{4}$$

Fig. 3 Workflow

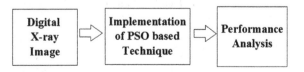

$K(i, j)$ is the enhancement function and defined as

$$S(i, j) = \frac{k.D}{\sigma(i, j) + b} \tag{5}$$

$\sigma(i, j)$ is the standard deviation of the window with size $n \times n$ and defined as

$$\sigma(i, j) = \sqrt{\frac{1}{n \times n}(\sum_{i=0}^{n}\sum_{j=0}^{n} y(i, j) - p(i, j)^2} \tag{6}$$

D is the global mean and defined as

$$R = \frac{1}{n \times n} \sum_{i=0}^{M-1}\sum_{j=0}^{N-1} y(i, j) \tag{7}$$

Finally, the transformation function is defined as

$$x(i, j) = \frac{k.R}{\sigma(i, j) + b}[y(i, j) - c \times p(i, j)] + p(i, j)^a \tag{8}$$

The value of the four parameters (a, b, c, k) are being optimized through parameter tuning are described in Sects. 4.2 and 4.3.

4.2 Objective Function

It is a measurement of quality of an image without any human intervention. While constructing the objective function, we have considered three properties of the enhanced images:

- A good contrast enhanced image has more number of edges than the original input image.
- An enhanced image has more intensity value than its edge.
- Entropy: This reveals the information content of an image. If the intensity distribution of the image is uniform, it should have high.

Hence, the objective function for this purpose is

$$F(I_e) = \log(\log E(I_s)) \times n_{edgles} \times H(I_e) \tag{9}$$

where I_e, I_s are the enhanced image and edge detected image, respectively. $E(I_s)$ is the sum of pixel intensity of edge detected image I_s. n_{edgles} is the number of edge pixels that have intensity value more than threshold value used in Sobel edge detector; in this case, we have used Otsu's global thresholding technique. Entropy of the image is calculated as

$$H(I_e) = -\sum_{i=0}^{255} e_i \tag{10}$$

$e_i = h_i \log_2 h_i$ if $h_i \neq 0$; otherwise, $e_i = 0$. h_i is the probability of a intensity value to be in the image. I_e is the enhanced image of an input image $I \cdot I_s$ is the edge image that can be defined using the kernel function described in Eqs. 12 and 13 and can be defined as

$$I_s = \sqrt{\delta m I_e(i, j)^2 + \delta n I_e(i, j)^2} \tag{11}$$

$$\delta m I_e = g I_e(i+1, j-1) + 2g I_e(i+1, j) + g I_e(i+1, j+1) - g I_e(i-1, j-1) - 2g I_e(i-1, j) - g I_e(i-1, j+1) \tag{12}$$

$$\delta n I_e = g I_e(i-1, j+1) + 2g I_e(i, j+1) + g I_e(i+1, j+1) - g I_e(i-1, j-1) - 2g I_e(i, j-1) - g I_e(i+1, j-1) \tag{13}$$

4.3   Particle Swarm Optimization

In this study, we have used PSO to obtain the optimal intensity value of the enhanced image. This technique is implemented based on the principle of swarm intelligence family. There are numerous techniques in swarm intelligence umbrella like genetic algorithm (GA), artificial bee colony (ABC), ant colony optimization (ACO), etc. GA computes the optimal solution using crossover and mutation. ACO and artificial bee colony (ABC) compute the optimal solution using food searching behavior of ant and bee swarm, respectively, whereas in PSO, it works based on the swarm behavior of bird [15]. In this study, we have used PSO to enhance the contrast of the image. The results clearly show that in case of X-ray image enhancement, PSO behaves better than the other members of swarm family. Each solution in this algorithm is known as particle. These particles are moved through the problem iteratively. Considering both local and global information of a low contrast image, the transformation function is used to enhance the gray value of the image using the objective functions. a, b, c, k are four parameters that are responsible to get the optimum result using fitness value, which is a parameter of good quality images. These values are computed through PSO mechanism. For a P number of particles, the four parameters are initialized within a range and equivalent to random velocities calculated as

$$v_i^{t+1} = w^t v_i^t + c_1 r_1 (pbest_i^t - X_i^t) + c_2 r_2 (gbest^t - X_i^t) \tag{14}$$

$$X_i^{t+1} = X_i^t + v_i^{t+1} \tag{15}$$

where X_i^t and v_i^t are the positions and velocity at time t, and W^t is the inertia weight which can be defined as

$$W^t = W_{max} \frac{W_{max} - W_{min}}{t_{max}} \times t. \tag{16}$$

Each vector for each particle has four components a, b, c, k. From this fitness value, the g_{best} and p_{best} known as local and global best values are calculated to direct the particle in a proper direction to obtain optimum solutions. Hence, new positions of particle lead to a better enhanced image of the histogram.

The contrast of the image is enhanced through the transformation function, which involves both global and local intensity values of the image. For each particle N, the four parameters represent the four components of the position vector. The parameter ranges defined in Sect. 4.3 represent the random velocity. This section described about the optimized velocity required in contrast enhancement. The quality of the image is measured through the objective function which is none other than the fitness of the particle. The $gbest$ and $pbest$ values help to move the particle toward the best position. In each iteration, a set of new particles have been generated along with its $gbest$ and $pbest$ values using the objective functions. The process is terminated when the image is enhanced through its $gbest$ value.

4.3.1 Parameter Tuning

PSO is a parameter-dependent algorithm. Gray value enhancement of benchmark image using PSO is introduced in [14], and they have used the value of four parameters used in this PSO-based enhancement method which lies between $a \in [0, 1.5]$, $b \in [0, 0.5]$, $c \in [0, 1]$, $k \in [0.5, 1.5]$.

However, there are some differences in properties of normal images with those of the medical images. All the images generated from same X-ray scanner machine have the same intensity property. Hence, to calculate the optimal value of each parameters, we have executed Algorithm 1, which yields the fitness value described in Eq. 9 in a range of $a \in [0, 2]$, $b \in [0, 1.2]$, $c \in [0, 2]$, $k \in [0, 2]$. The value of b changes the output image drastically and is extremely sensitive. Figure 6 clearly shows that with very high and very low value of the above-mentioned parameter b generates a binary image. The plots shown in Fig. 4a, b, c are three-dimensional line plots, where fitness value is plotted against any two variables keeping the third variable constant. These three plots yield a pair of value for each of the variables. Finally, the optimum values for the variables are calculated by averaging it.

Algorithm 1 PSO in Contrast Enhancement

Input: Low Contrast Image
Output: Contrast Enhanced Image

1: Define N number of particles with m dimension
2: values of parameter a,b,c,k are initialized
3: **for** Each Particle $i = 1$ to N **do**
4:    gbest and pbest of each particles are calculated using Equation 9.
5:    The inertia weight (W) is calculated as

$$W^t = W_{max} - \frac{W_{max} - W_{min}}{t_{max}} \times t. \qquad (17)$$

6:    **while** (gbest$\neq$ pbest)& $W^t > W_{min}$ **do**
7:        Calculate fitness value using the objective function described in 9
8:        **if** $T(I_e)_i < T(pbest_i)$ **then**
9:            $pbest_i = N_i$
10:        **end if**
11:        **if** $T(I_e)_i > T(gbest_i)$ **then**
12:            $gbest_i = N_i$
13:        **end if**
14:        Update velocity and position using Equation 14 and 15
15:
16:        Enhanced Image is generated using Equation 8
17:    **end while**
18: **end for**

Figure 4a shows that value of (a, k) is $(0.7, 0.9)$
Figure 4b shows that value of (k, c) is $(0.7, 0.8)$
Figure 4c shows that value of (a, c) is $(0.9, 0.9)$
Hence, $a = 0.8$, $k = 0.85$, $c = 0.9$. The window size is determined according to the base exposure of the image. Higher the window size, the enhanced image will be more smoother, i.e., more articulation will be there in the output image. Lesser the window size, the output image produced will be toward a binary image, and is used in case of moderate exposure images. The window size needs to be optimized to enhance any presence of any abnormal entity. However, in case of any low exposure image, the window size needs to be large, to enhance the image and to provide more accurate estimation of any presence of any external entity and also to provide a better fitness value of the enhanced image. Figure 5 shows that at $n \leq 30$ and $n \geq 50$ yields a binary image.

Fig. 4 3D Plot for
parameter tuning

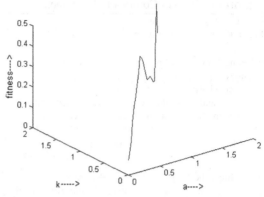

(a) variation of Fitness function with respect to a,k

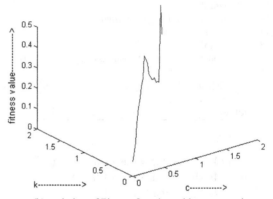

(b) variation of Fitness function with respect to k,c

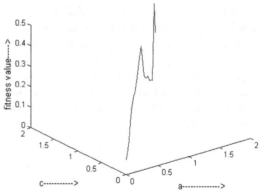

(c) variation of Fitness function with respect to a,c

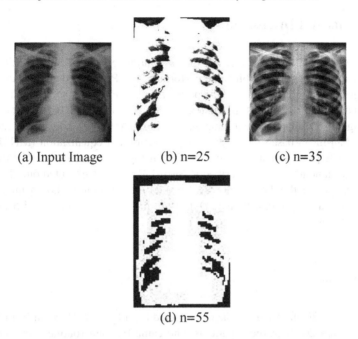

(a) Input Image   (b) n=25   (c) n=35

(d) n=55

Fig. 5 Output based on window size

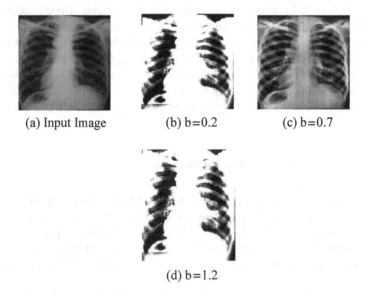

(a) Input Image   (b) b=0.2   (c) b=0.7

(d) b=1.2

Fig. 6 Result of parameter tuning for variable b

5 Results and Discussion

In this section, we display the result of our contrast enhancement algorithm over digital X-ray images. To establish the performance of PSO-based contrast enhancement algorithm, we have analyzed the performance of PSO over some standard contrast enhancement techniques like histogram equalization (HE) [16], linear contrast stretching (LCS), adaptive histogram equalization (AHE) [7], adjustment function [16], median-mean-based sub-image-clipped histogram equalization (MNSICHE) [9], BPFDHE [11], nonparametric modified histogram equalization (NHMC) [8] using some standard metric described in Sect. 5.3. We have executed our algorithm over all images of the dataset, however results of 5 randomly chosen images are shown below among them, Image ID 1–3 are high contrast and 4 and 5 are low contrast images.

5.1 Dataset

We have used 220 chest X-ray images collected from Peerless Hospital from March 2015 to March 2017 for those patients who came for their routine check up with different types of lung diseases for contrast enhancement [17]. The patient edge lies between 30–75 which consists of both male and female patients. Among 220 patients, 157 patients are male and rest are female. In conventional X-ray imaging procedure, the X-ray beam passes through the human body the shadow of hard tissue as shadow on the film. This film is then processed and printed to get the picture of the organ. In this study, we have used CR type of digital X-ray images where a sensor is placed behind the patient instead of placing X-ray films in conventional radiography. This sensor is next attached to the computer to create the digital radiography [18].

5.2 Result of Annotation Process

Three experienced radiologist have been found 300 abnormalities from 120 chest radiographs, among them, 257 cases are marked as nodule by at least one radiologists and 227 cases marked as nodule by all radiologists. Although, Biopsy and cytology report says that there present 252 nodules and among them 193 cases are malignant rest are benign among all the malignant cases 70 are responsible for metastatic lung cancer.

The type of nodule is further categorized into five classes, namely obvious, relatively obvious, subtle, very subtle, and extremely subtle. There are 86 nodules that are very subtle, 31 nodules are extremely subtle, 48 nodules are subtle, 64 nodules are obvious, and the rest of 33 nodules are relatively obvious.

Anatomy of human lung reveals that there are five lobes and generation in different positions has different chances of malignancy. There are 88 nodules that are in left lower lobe, 72 nodules are in left upper lobe, 10 nodules are in the right middle lobe, 56 nodules are in right lower lobe, and 26 nodules are in right upper lobe.

5.3 Evaluation Metrics

- Peak signal-to-noise ratio(PSNR): PSNR can be defined as the ratio between the maximum power of the signal and the power of the corrupting noise, which measures the peak error. Higher the PSNR value indicates the higher quality of images. Mathematically, $PSNR$ can be defined as:

$$PSNR = 20 \log_{10}(\frac{255}{RMSE})$$ (18)

- Mean square error:

$$MSE = \sqrt{\frac{1}{n} \sum_{i=1}^{n} (x - y)^2}$$ (19)

where x is the output image and y is the reference image.
- Mean absolute brightness error (MABE): It is the measurement of absolute intensity value difference between input image and enhanced image

$$D = \|\mu_A - \mu_B\|.$$ (20)

An algorithm generating less value is considered as good one.
- Detailed variance:
- Background variance:

Figure 7 exhibits the visual verification result of input chest X-ray images. The input image in Fig. 7 is much lower in contrast than that of the input images of Fig. 8. The output of these figures along with their histogram clearly shows that PSO works better in case of comparatively very low contrast X-ray images even it suppresses the ribs which are not our region of interest. However in comparatively high contrast images, it does not score very high. Tables 2, 4, 1 clearly show that PSO-based image enhancement algorithm works better than the modified histogram equalization methods. However, the visual verification result always does not give the guarantee of perfect results (Figs. 9 and 10).

According to the definitions of fitness value, PSNR, Detailed Variance, Background Variance and MABE, MSE a good contrast image has higher fitness value, PSNR, detailed variance and background variance and least value in case of MABE and MSE. The comparative analysis of this metric is described in Tables 1, 2, 3, 4, 5, 6 and Figs. 9, 8, 11, 12, 13, 14.

Fig. 7 Visual verification of PSO over other algorithm for low contrast image

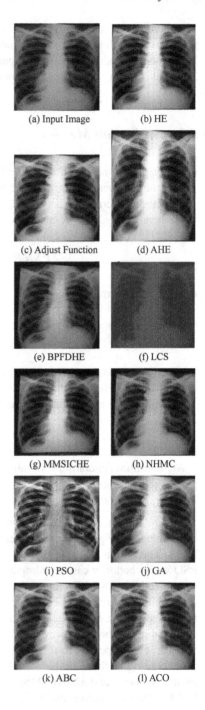

(a) Input Image   (b) HE

(c) Adjust Function   (d) AHE

(e) BPFDHE   (f) LCS

(g) MMSICHE   (h) NHMC

(i) PSO   (j) GA

(k) ABC   (l) ACO

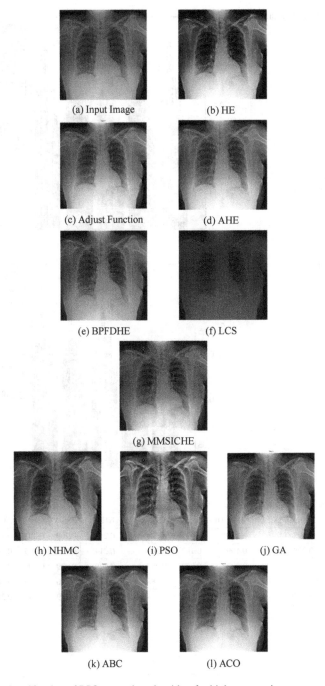

Fig. 8 Visual verification of PSO over other algorithm for high contrast images

Fig. 9 Background variance

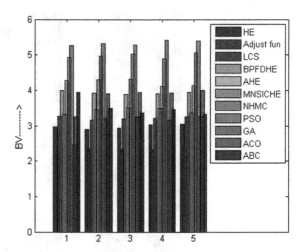

Fig. 10 Comparison of fitness value

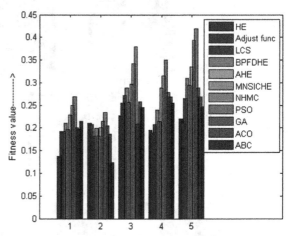

Table 1 Fitness function value of different algorithms

Image	HE	Adjust	LCS	BPFDHE	AHE	MNSICHE	NHMC	PSO	GA	ACO	ABC
ID1	0.138	0.192	0.192	0.210	0.197	0.229	0.250	0.269	0.201	0.198	0.215
ID2	0.210	0.208	0.182	0.199	0.182	0.201	0.214	0.234	0.205	0.187	0.123
ID3	0.227	0.255	0.272	0.289	0.257	0.297	0.342	0.379	0.209	0.258	0.245
ID4	0.195	0.189	0.207	0.239	0.214	0.288	0.315	0.350	0.279	0.269	0.255
ID5	0.220	0.215	0.265	0.310	0.294	0.3347	0.393	0.419	0.289	0.269	0.247

Table 2 Comparison of background variance

Image	HE	Adjust	LCS	BPFDHE	AHE	MNSICHE	NHMC	PSO	GA	ACO	ABC
ID1	2.97	2.28	3.27	3.99	3.32	4.27	4.92	5.27	2.46	3.25	3.94
ID2	2.89	2.36	3.15	3.92	3.45	4.29	4.96	5.32	3.21	3.91	3.49
ID3	2.94	2.34	3.19	3.89	3.52	4.31	5.02	5.29	3.25	3.94	3.37
ID4	3.02	2.32	3.21	3.91	3.49	4.11	4.89	5.41	3.15	3.92	3.45
ID5	3.05	2.46	3.25	3.94	3.37	4.13	5.06	5.39	3.27	3.99	3.32

Table 3 Detailed variance generated by different algorithms

Image	HE	Adjust	LCS	BPFDHE	AHE	MNSICHE	NHMC	PSO	GA	ACO	ABC
ID1	0.1873	0.1507	0.2030	0.2870	0.2507	0.3005	0.3220	0.3350	0.2356	0.2545	0.2356
ID2	0.1806	0.1512	0.2005	0.2805	0.2520	0.2997	0.3216	0.3370	0.1245	0.1986	0.2109
ID3	0.1897	0.1577	0.2078	0.2890	0.2594	0.3047	0.3205	0.3397	0.2589	0.2548	0.2989
ID4	0.1809	0.1502	0.2009	0.2820	0.2509	0.3020	0.3195	0.3398	0.2256	0.2548	0.2378
ID5	0.1825	0.1508	0.2083	0.2837	0.2610	0.3017	0.3189	0.3390	0.2356	0.2454	0.2545

Table 4 Mean absolute brightness error

Image	HE	Adjust	LCS	BPFDHE	AHE	MNSICHE	NHMC	PSO	GA	ACO	ABC
ID1	22.17	27.33	24.67	16.23	20.13	15.51	12.72	9.44	10.23	11.89	12.56
ID2	22.52	27.45	23.78	15.52	19.89	1.23	13.23	10.15	10.02	11.89	12.56
ID3	22.12	28.06	23.52	16.09	20.05	15.21	12.99	9.74	12.23	11.56	12.89
ID4	21.86	27.96	24.26	15.89	19.97	15.04	12.85	9.56	13.26	15.30	11.56
ID5	22.05	27.69	23.56	15.75	19.56	15.53	12.89	9.45	13.02	12.56	15.23

Table 5 Peak signal-to-noise ratio

Image	HE	Adjust	LCS	BPFDHE	AHE	MNSICHE	NHMC	PSO	GA	ACO	ABC
ID1	42.17	47.33	54.67	46.23	50.13	45.51	42.72	59.14	54.12	55.45	57.89
ID2	42.52	47.45	53.78	45.52	49.89	4.23	43.23	60.25	53.93	55.01	56.89
ID3	42.12	48.06	53.52	46.09	50.05	45.21	42.99	63.74	54.21	54.69	56.05
ID4	41.86	47.96	54.26	45.89	49.97	45.04	42.85	62.56	54.01	55.23	57.09
ID5	42.05	47.69	53.56	45.75	49.56	45.53	42.89	59.45	53.22	54.95	57.22

Table 6 Mean square error

Image	HE	Adjust	LCS	BPFDHE	AHE	MNSICHE	NHMC	PSO	GA	ACO	ABC
ID1	61.23	65.12	60.23	66.23	55.13	41.51	42.72	19.14	54.12	25.45	37.89
ID2	62.78	65.45	60.56	65.52	55.89	41.23	43.23	19.25	53.93	25.01	36.89
ID3	61.95	65.89	60.45	66.09	56.05	42.21	42.99	20.74	54.21	24.69	36.05
ID4	62.12	64.89	60.78	65.89	56.97	43.04	42.85	19.56	54.01	25.23	37.09
ID5	63.12	67.23	60.12	65.75	56.56	42.53	42.89	19.45	23.22	54.95	37.22

Fig. 11 Comparison of
PSNR

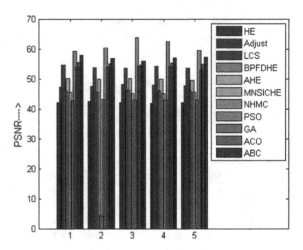

Fig. 12 Comparison of
entropy

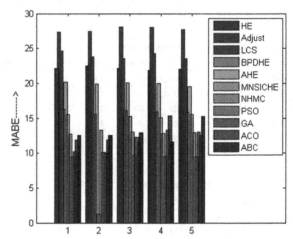

Table 7 Result of cross-validation

ID	FitnessValue	BV	DV	PSNR	MSE	MABE
Image1	0.269	5.27	0.3350	98.56	0	0
Image2	0.234	5.32	0.3370	98.78	0	0
Image3	0.379	5.29	0.3397	98.95	0	0
Image4	0.350	5.41	0.3398	99.23	0	0
Image5	0.419	5.39	0.3390	99.04	0	0

To ensure that the output enhanced images are optimized, we have passed the
output image as input image in algorithm to calculate difference metrics as described
in Sect. 5.3. The evaluated metric values are shown in Table 7.

Fig. 13 Comparison of mean square error

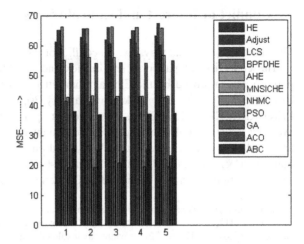

Fig. 14 Comparison of detailed variance

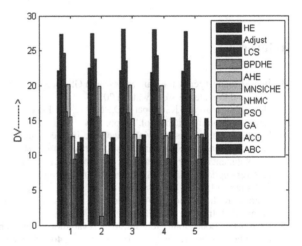

6 Conclusion

Contrast enhancement of medical images is quite a challenging task in medical image analysis. In this paper, we have applied particle swarm optimization over digital X-ray images. The results in comparison with the other existing techniques are quite satisfactory. In future, we will apply more accurate objective function and/or any other bio-inspired meta-heuristic techniques rather than swarm intelligence to get better enhanced images.

7 Conflict of Interest and Ethical Issues

The authors declare that they have no conflict of interest.

The collection of patient scans images and pathological report for research purpose was approved by the Ethical Committee of Peerless Hospital and B. K. Roy Research Centre Ltd, Kolkata on January 30, 2015 with reference number: PH&BKRRCCREC/1576/2015. This study is approved till January 2020. However, the issue of publishing the collected patient images as a public dataset was approved by the same ethical committee dated on August 4, 2016, with reference number:PH&BKRRCCREC/2063/2016, abide by all the ethical guidelines of a retrospective study generated by the Indian Council of Medical Research (ICMR).

Acknowledgements We are thankful to Center of Excellence in Systems Biology and Biomedical Engineering (TEQIP II) of University of Calcutta for providing the financial support for this research and Peerless Hospitex Hospital for providing their valuable image dataset.

References

1. International Agency for Cancer Research Globocan 2012 Estimated Cancer Incidence, mortality and Prevalence Worldwide in 2012
2. Heath, M., Bowyer, K., Kopans, D., Moore, R., Kegelmeyer Jr P.: The digital database for screening mammography, vol. 58, pp. 27,96
3. Shiraishi, Junji, Katsuragawa, Shigehiko, Ikezoe, Junpei, Matsumoto, Tsuneo, Kobayashi, Takeshi, Komatsu, Ken-ichi, Matsui, Mitate, Fujita, Hiroshi, Kodera, Yoshie, Doi, Kunio: Development of a digital image database for chest radiographs with and without a lung nodule: receiver operating characteristic analysis of radiologists' detection of pulmonary nodules. Am. J. Roentgenol. **174**(1), 71–74 (2000)
4. Sherrier, R.H., Johnson, G.A.: Regionally adaptive histogram equalization of the chest. IEEE Trans. Med. Imaging, **6**(1), 1–7 (1987)
5. Altas, Irfan, Louis, John, Belward, John: A variational approach to the radiometric enhancement of digital imagery. IEEE Trans. Image Process. **4**(6), 845–849 (1995)
6. Chen, S.D., Ramli, A.R.: Contrast enhancement using recursive mean-separate histogram equalization for scalable brightness preservation. IEEE Trans. Consum. Electron. **49**(4), 1301–1309 (2003)
7. Zuiderveld, K.: Contrast limited adaptive histogram equalization, Graphics gems IV, pp. 474–485. Academic Press Professional, Inc (1994)
8. Poddar, S., Tewary, S., Sharma, D., Karar, V., Ghosh, A., Pal, S.K.: Non-parametric modified histogram equalisation for contrast enhancement. IET Image Process. **7**(7), 641–652 (2013)
9. Singh, Kuldeep, Kapoor, Rajiv: Image enhancement via median-mean based sub-image-clipped histogram equalization. Opt. Int. J. Light Electron Opt. **125**(17), 4646–4951 (2014)
10. Xue, Q.: Enhancement of medical images in the shearlet domain. In: Computer Science and Network Technology (ICCSNT), 2013 3rd International Conference on, pp. 235–238. IEEE (2013) (October)
11. Sheet, Debdoot, Garud, Hrushikesh, Suveer, Amit, Mahadevappa, Manjunatha, Chatterjee, Jyotirmoy: Brightness preserving dynamic fuzzy histogram equalization. IEEE Trans. Consum. Electron. **56**(4), 2475–2481 (2010)
12. Yang, H.Y., Lee, Y.C., Fan, Y.C., Taso, H.W.: A novel algorithm of local contrast enhancement for medical image. In: Nuclear Science Symposium Conference Record, 2007. NSS'07, vol. 5, pp. 3951-3954. IEEE (2007) (October)

13. Hashemi, S., Kiani, S., Noroozi, N., Moghaddam, M.E.: An image contrast enhancement method based on genetic algorithm. Pattern Recognit. Lett. **31**(13), 1816–1824 (2010)
14. Gorai, A., Ghosh, A.: Gray-level image enhancement by particle swarm optimization. In: World Congress on Nature & Biologically Inspired Computing, 2009. NaBIC 2009, pp. 72–77 (2009)
15. Eberhart, R.C., Kennedy, J.: A new optimizer using particle swarm theory. In: Proceedings of the Sixth International Symposium on Micro Machine and Human Science, vol. 1, pp. 39–43 (1995)
16. Gonzalez, R.C., Woods, R.E.: Digital image processing, 3rd edn. Tata MacgrawHill (2008)
17. http://www.coe.cuteqip.net/coepeerlessxray.php
18. Wendt, R.: The Mathematics of Medical Imaging: A Beginner's Guide, pp. 1987–1987 (2010)
19. Stetson, P.F., Sommer, F.G., Macovski, A.: Lesion contrast enhancement in medical ultrasound imaging. IEEE Trans. Med. Imaging **16**(4), 416–425 (1997)
20. Zhu, H., Chan, F.H., Lam, F.K.: Image contrast enhancement by constrained local histogram equalization. Comput. Vis. Image Underst. **73**(2), 281–290 (1999)
21. Al-Manea, A., El-Zaart, A.: Contrast enhancement of MRI images. In: Ibrahim, F., Osman, N.A.A., Usman, J., Kadri, N.A. (eds) 3rd Kuala Lumpur International Conference on Biomedical Engineering 2006. IFMBE Proceedings, vol 15. Springer, Berlin, Heidelberg (2007)
22. Knopp, M.V., Giesel, F.L., Marcos, H., von Tengg-Kobligk, H., Choyke, P.: Dynamic contrast-enhanced magnetic resonance imaging in oncology. Top. Magn. Reson. Imaging **12**(4), 301–308 (2001)

Ultrasound Medical Image Deblurring and Denoising Method Using Variational Model on CUDA

Biswajit Biswas, Biplab Kanti Sen and Kashi Nath Dey

Abstract This paper introduces a new variational model on CUDA platform for the restoration (deblurring and denoising) of ultrasound image degraded by additive Gaussian noise and blur effect. In the deblurring step, we apply an inverse algorithm with the fast transform approach. In the denoising step, a total variational model (TVM) using second-order partial anisotropic diffusion equations is used. A unique and stable solution for the proposed model is presented in terms of the Euler–Lagrange equation. Later, an accurate numerical approximation is constituted by the finite-difference-based discretization technique and the parameter dependence of the proposed model is also described. To achieve better acceleration with satisfactory performance, the proposed algorithm is properly devised on the CUDA GPU and compared with a sequential execution of the multicore CPU system. Experimental results and quantitative analysis show that our algorithm is efficient to restore the ultrasound image compared to the state-of-the-art restoration methods.

Keywords Ultrasound image denoising · Partial differential equation
Total variation · Euler–Lagrange equation · Computation Unified Device
Architecture (CUDA) · Graphics Processing Units (GPU) · Signal-to-noise ratio

1 Introduction

In the last two decades, advancement in ultrasonic medical technology effectively generated a number of imaging techniques that are extensively practiced

B. Biswas · B. K. Sen (✉) · K. N. Dey (✉)
Department of Computer Science and Engineering, University of Calcutta, Kolkata,
West Bengal, India
e-mail: bksen.cu@gmail.com

K. N. Dey
e-mail: kndey55@gmail.com

B. Biswas
e-mail: biswajit.cu.08@gmail.com

© Springer Nature Singapore Pte Ltd. 2018   95
R. Chaki et al. (eds.), *Advanced Computing and Systems for Security*,
Advances in Intelligent Systems and Computing 666,
https://doi.org/10.1007/978-981-10-8180-4_6

by both researchers and clinicians for diagnostic ultrasound analysis. Each ultrasonic image sequence describes detailed anatomical and functional information about the anatomy of living tissues [1, 2]. Performance and efficiency of sonogram diagnostics, such as pixel-based tissue classification, extraction of organic shape, location, and tissue boundaries, are influenced by the amount of noise exists at the acquisition phase. The main cause of this kind of degradation is random thermal noise injecting into the sonogram data at the time of image acquisition [3, 4]. In recent, many researchers have addressed about some special random noise, such as multiplicative noise, impulse noise, and Poisson noise [1, 2, 4] in medical image processing. In this work, we are dealing with the restoring issue under the additive Gaussian noise with TVM. The image restoration problem under the additive noise like multiplicative noise is a challenging task. However, several approaches have been suggested to address the additive noise removal problem, for example, variational methods [2, 5, 12], filtering techniques [1, 2], and statistical method [1, 2]. Among the variational methods, total variation (TV)-based models are considerably effective and efficient for restoring the noisy image and better ability to preserve image details (e.g., edges, textures) [1, 2, 6]. Typically, image restoration is the process of restoring the image information around the noisy zone [1–3, 7]. Mostly, those restoration techniques use the partial differential equation (PDE) and total variational functional (TVF) [1, 2, 8]. A simple variational approach for image restoration method based on the Mumford-Shah functional has been developed in [2, 9], and another method using the Mumford-Shah algorithm found in [1, 2, 4], which solves restoration problem by means of level lines with minimal curvature in the image plane. An effective variational model for image restoring was introduced in [1, 3, 10] where the total variation (TV) restoration model was proposed. Those models use an Euler–Lagrange equation and anisotropic diffusion, based on the strength of the isophotes [1, 3, 11]. Recently, the TV restoration model has been extended and considerably improved the reconstruction methods, such as TV restoration with Split Bregman [4, 10, 12]. Similarly, we found other state-of-the-art variational restoration models based on TV regularization [2, 7, 11] and wavelet restoration TV [1, 3, 4] in the literature. In addition, a number of powerful image restoration models are recently developed by using the kriging interpolation approach [1, 2, 8]. Many research works have been published which are more efficient for noise removal [1, 10, 12]. Most popular techniques, such as bilateral filtering for denoising [2, 3], wavelet-based denoising [1, 2], and TV-based variational [10, 11] approaches, are successfully practiced in digital image processing. The first total variation-based noise removal model based on the constrained optimization approach with two Lagrange multipliers was introduced in [11, 12]. To overcome the drawback, the total generalized variation (TGV) regularizer integrated with an additional term known as the data fidelity term [1, 6–8]. From all the fact, all the different TVM designed for large-scale images, the restoration process is found to be computationally expensive with higher execution cost.

A lot of image processing algorithms (e.g., image denoising) has been successfully implemented in many accelerators, such as graphics processing units (GPUs) [1, 2, 13], field programmable gate arrays [1, 2], and cell broadband engines [1, 13]. The GPU is one of the most powerful hardware devices among them and extensively used in high computational tasks, where it contains an extensive higher memory bandwidth than the CPU. For instance, the latest Maxwell architecture provides 336 GB/s of memory bandwidth which is higher than the double data rate of the latest generation memory system. Both memory-bound and compute-bound algorithms have been efficiently accelerated as parallel to GPUs [13, 14]. Image restoration is a memory-bound operation as well as a compute-bound operation since we consider that the GPU is a perfect accelerator for the large-scale image restoration problems [13, 14].

This paper proposes a novel partial differential equation (PDE) based on total variation (TV) for image restoration. The proposed restoration scheme is derived on the basis of an anisotropic diffusion PDE-based variational model that efficiently denoises and deblurs the degraded sonogram images. In our nonlinear diffusive PDE model, we have used trigonometric function that deals with the anisotropic properties of an image (edge, corners, texture, etc.) and measure the orientation of those properties which helps to produce more detailed image rather than the state-of-the-art methods. It also performs deblurring and denoising simultaneously, which is also a unique contribution of our work. A unique and uniform solution to the proposed variational model is formed in terms of the Euler–Lagrange equation. Later, an accurate numerical approximation method is developed by finite-difference-based discretization approach and the parameter dependence of the proposed model is also investigated. Next, the implementation of the proposed model is done on the NVIDIA Quadro K-420. Experimental results show that the proposed model is outstanding with compared to the state-of-the-art restoration methods in terms of both quality and quantity.

The rest of this paper is organized as follows: The proposed variational model with the nonlinear diffusion scheme is described in Sect. 2. A stable numerical approximation scheme is introduced in Sect. 3. Section 4 presents the CUDA implementation in detail. Section 5 describes the experimental results in short. Finally, the concluding remarks are summarized in Sect. 6.

2 The Proposed Algorithm

The minimization of TV with an adaptive novel energy functional $\mathbf{E}$ is designed for the proposed model. Hence, the enhanced restored image denotes as $\mathbf{u}_r$ is constructed as an outcome of minimization of the $\mathbf{E}$ [1, 2, 4, 6–8]. Let $\mathbf{u}$ be the original image, and it should be restored from the observed image $\mathbf{u}_0$. For a given image $\mathbf{u} \in \mathbf{L}^2(\Omega)$, with $\Omega \subset \mathbf{R}^2$ an open and bounded domain, the TV-based models are uniformly written as follows [1, 2, 4, 6–8]:

$$\mathbf{u}_r = \underset{\mathbf{u}}{\operatorname{argmin}} \int_\Omega \left(\frac{\beta}{2} \|\mathbf{H}\mathbf{u} - \mathbf{u}_0\|_2^2 + \frac{\lambda}{2} \|\mathbf{u} - \mathbf{u}_0\|_2^2 + \alpha \|\nabla u\|_1 \right) \partial\Omega \qquad (1)$$

where the first term is the blurring components of $\mathbf{u}$, on the other hand, $\|\mathbf{u} - \mathbf{u}_0\|_2^2$ is the data fidelity terms, the $\alpha \|\nabla u\|_1$ is total variation, and β, λ, α are the regularization parameters. Matrix H is a blur matrix that contains Fourier basis.

TVM of the energy functional $\mathbf{E}$ can be formulated as follows [1, 2, 4, 6–8]:

$$\mathbf{E}(\mathbf{u}) = \underset{\Omega}{\arg\min} \int_{\Omega} \left(\frac{\beta}{2} \|\mathbf{Hu} - \mathbf{u}_0\|_2^2 + \frac{\lambda}{2} \|\mathbf{u} - \mathbf{u}_0\|_2^2 + \alpha \|\nabla u\|_1 \right) \partial\Omega \quad (2)$$

where the first term is the regularization term which controls the quality of the image $\mathbf{u}$. ∇u denotes the Laplacian of $\mathbf{u}$ and $\|\nabla u\| = \sqrt{|\mathbf{u}|_{xx}^2 + |\mathbf{u}|_{yy}^2 + \epsilon}$, and the parameter $\epsilon > 0$ is introduced to avoid the singularity. The second term is the data fidelity which quantifies the violation of the relation between $\mathbf{u}$ and the observed image $\mathbf{u}_0$, and $\alpha, \beta \in (0, 1)$ are the regularization parameters [1, 2, 4, 6–8]. In this work, we choose α and β adaptively and it is estimated by the gradient function of the evolving image $\mathbf{u}$, which is expressed as follows:

$$\alpha = \left(\frac{\|\nabla u\|_1}{\alpha_1 \|\nabla u\|_1 + \alpha_2} \right), \beta = \left(\frac{\|\nabla u\|_2}{1 + \alpha_1 \|\nabla u\|_2 + \alpha_2} \right) \quad (3)$$

where $\alpha_1 > 0$, $\alpha_2 > 0$ are two arbitrary parameters. In our experiment, we found satisfactory result when $\alpha = 0.1$ and $\beta = 0.1$. To determine an exact solution of the proposed model Eq. (2), the iterative algorithm starts with two minimization subproblems and can be expressed as follows [1, 2, 4, 6–8]:

(1) Denoising step. For $\mathbf{u}_0$ fixed, find the solutions of $\mathbf{u}$

$$\mathbf{u}^{k+1} = \underset{u \in \Omega}{\arg\min} \, \alpha \int_{\Omega} \|\nabla u\|_1 \, \partial\Omega + \frac{\lambda}{2} \int_{\Omega} \|\mathbf{u}^k - \mathbf{u}_0\|_2^2 \partial\Omega \quad (4)$$

(2) Deblurring step. For $\mathbf{u}_0$ fixed, find the solution of $\mathbf{u}$

$$\mathbf{u}^{k+1} = \underset{u}{\arg\min} \, \frac{\beta}{2} \|\mathbf{Hu}^{k+1} - \mathbf{u}_0\|_2^2 + \frac{\lambda}{2} \|\mathbf{u}^{k+1} - \mathbf{u}^k\|_2^2 \quad (5)$$

Now, we designed the corresponding algorithms for Eqs. (4) and (5), separately. First, for Eq. (4), we note $E(x, y, u, u_x, u_y) = \alpha(\|\nabla u\|)\lambda(\mathbf{u}^k - \mathbf{u}_0)^2$, where $u_x = \frac{\partial u}{\partial x}$; by applying the Euler–Lagrange equation corresponding to the variational model [1, 2, 4, 6–8], we obtain

$$\frac{\partial E}{\partial u} - \frac{\partial}{\partial x} \frac{\partial E}{\partial u_x} - \frac{\partial}{\partial y} \frac{\partial E}{\partial u_y} = 0 \quad (6)$$

that leads to—$\lambda(\mathbf{u}^k - \mathbf{u}_0) - \frac{\partial}{\partial x}\left(\alpha \mathbf{u}^k(\|\nabla u\|)\frac{2u_x}{\|\nabla u\|}\right) - \frac{\partial}{\partial y}\left(\alpha \mathbf{u}(\|\nabla u\|)\frac{2u_y}{\|\nabla u\|}\right) = 0$ that is equivalent to

$$\lambda \left(\mathbf{u}^k - \mathbf{u}_0 \right) - 2\alpha \nabla \cdot \left(\frac{(\nabla u)}{\|\nabla u\|} \nabla u \right) = 0 \tag{7}$$

Then, one applies the gradient descent [1, 2, 4] and the following nonlinear diffusion PDE model [4–8] is obtained from Eq. (7):

$$\begin{cases} \frac{\partial u}{\partial t} = 2\alpha \nabla \cdot \left(\frac{(\nabla u)}{\|\nabla u\|} \nabla u \right) - \lambda \left(\mathbf{u} - \mathbf{u}_0 \right) \\ u\left(0, x, y \right) = u_0 \\ u\left(t, x, y \right) = 0 \text{ on, } \partial \Omega \backslash \Gamma \end{cases} \tag{8}$$

Second, for Eq. (5), its corresponding Euler–Lagrange equation is formulated as follows [4, 6–8]:

$$\beta \mathbf{H}^T \left(\mathbf{H} \mathbf{u}^{k+1} - \mathbf{u}_0 \right) + \lambda \left(\mathbf{u}^{k+1} - \mathbf{u}^k \right) = 0 \tag{9}$$

Therefore, we have

$$\left(\lambda \mathbf{I} + \beta \mathbf{H}^T \mathbf{H} \right) \mathbf{u}^{k+1} = \left(\lambda \mathbf{u}^k + \mathbf{u}_0 \beta \mathbf{H}^T \right) \tag{10}$$

where $\lambda \mathbf{I}$ is a regularized term with identity matrix $\mathbf{I}$, and the $\left(\lambda \mathbf{I} + \beta \mathbf{H}^T \mathbf{H} \right)$ is an invertible matrix. After simplification, we obtain an approximate solution from Eq. (10) as follows:

$$\mathbf{u}^{k+1} = \left(\lambda \mathbf{I} + \beta \mathbf{H}^T \mathbf{H} \right)^{-1} \left(\lambda \mathbf{u}^k + \mathbf{u}_0 \beta \mathbf{H}^T \right) \tag{11}$$

where the matrix $\mathbf{H}$ is diagonalized by using the discrete fast transform [4, 6–8]. The result of minimization Eq. (2), representing the restored image, will be determined by solving with both Eqs. (8) and (11).

3 Numerical Approximation Scheme

A numerical approximation procedure relating the finite-difference-based discretization approach is carried out considering a space grid size of h and the time step Δt. The coordinates of space and time are quantized as follows [4, 5, 8]:

$$x = ih \approx i, \quad \forall i \in [1, \cdots, M]$$
$$y = jh, \approx j, \quad \forall j \in [1, \cdots, N]$$
$$t = k\Delta t, \quad \forall k \in [1, \cdots, K]$$

where h is horizontal space. For simplicity, we set the parameters $h = 1$ and $\Delta t = 1$. In the case of denoising component, we have $\nabla((\|\nabla u\|)\nabla u) = (\|\nabla u\|)\Delta u + \nabla((\|\nabla u\|)) \cdot \nabla u$, the Eq. (8) can be discretized as follows [1, 2, 4]:

$$\frac{\partial u}{\partial t} = 2\alpha((\|\nabla u\|)\Delta u + \nabla((\|\nabla u\|)) \cdot \nabla u) - \lambda(u - u_0) \tag{12}$$

The component relates to the Laplacian $(\|\nabla u\|)\Delta u$ operator, and it is approximated by using finite differences, as follows [4, 5, 8]:

$$D_{1_k}(i, j) = (\|\nabla u_k(i, j)\|)\nabla^2 u_k(i, j) \tag{13}$$

where

$$\nabla^2 u_k(i, j) = \frac{u_k(i+h, j) + u_k(i-h, j) + u_k(i, j+h) + u_k(i, j-h)}{h^2} - \frac{4u_k(i, j)}{h^2} \tag{14}$$

Also, we have

$$\nabla((\|\nabla u\|)) \cdot \nabla u = \left(\frac{\partial}{\partial x}\left(\sqrt{u_x^2 + u_y^2}\right), \frac{\partial}{\partial y}\left(\sqrt{u_x^2 + u_y^2}\right)\right) \cdot (u_x, u_y) \tag{15}$$

that leads to

$$\nabla((\|\nabla u\|)) \cdot \nabla u = \left(\sqrt{u_x^2 + u_y^2}\right)\frac{u_x^2 u_{xx} + 2u_x u_y u_{xy} + u_y^2 u_{yy}}{\sqrt{u_x^2 + u_y^2}} \tag{16}$$

We can simplify this formula, by performing some approximations. So, we may approximate $u_{xx} \approx u_{xy} \approx u_{yy}$, since we have found that second-order derivatives do not vary so much. Thus, Eq. (16) becomes

$$\nabla(\psi_u(\|\nabla u\|)) \cdot \nabla u \approx\approx \left(\sqrt{u_x^2 + u_y^2}\right)u_{xy}\left(u_x^2 + u_y^2\right) \tag{17}$$

So, we determine the second discretization that is

$$D_{2_k}(i, j) = \left(\sqrt{\left(\frac{\partial u_k(i, j)}{\partial i}\right)^2 + \left(\frac{\partial u_k(i, j)}{\partial j}\right)^2}\right) \cdot \frac{\partial^2 u_k(i, j)}{\partial i \partial j}\left(\frac{\partial u_k(i, j)}{\partial i} + \frac{\partial u_k(i, j)}{\partial j}\right) \tag{18}$$

where, by using the finite differences Eq. (17), we get

$$\begin{cases} \frac{\partial u_k(i,j)}{\partial i} + \frac{\partial u_k(i,j)}{\partial j} = \frac{u_k(i+h, j) - u_k(i-h, j) + u_k(i, j+h) - u_k(i, j-h)}{2h} \\ \frac{\partial^2 u_k(i,j)}{\partial i \partial j} = \\ \frac{u_k(i+h, j+h) - u_k(i+h, j-h) - u_k(i-h, j+h) - u_k(i-h, j-h) - 4u_k(i,j)}{4h^2} \end{cases}$$

On the other hand, in the case of deblurring components, Eq. (11) can be discretized as follows [4, 5, 8]:

$$\mathbf{u}\,(\mathbf{i}, \mathbf{j})^{k+1} = \left(\lambda \mathbf{I}\,(\mathbf{i}, \mathbf{j}) + \beta \mathbf{H}\,(\mathbf{i}, \mathbf{j})^T \mathbf{H}\,(\mathbf{i}, \mathbf{j})\right)^{-1} \left(\lambda \mathbf{u}\,(\mathbf{i}, \mathbf{j})^k + \mathbf{u}\,(\mathbf{i}, \mathbf{j})_0\,\beta \mathbf{H}\,(\mathbf{i}, \mathbf{j})^T\right) \quad (19)$$

From the above description, we obtain the approximated solution of the proposed model by using Eqs. (12) and (11) as follows:

$$\mathbf{u}_{k+1}\,(i, j) = \mathbf{u}_k\,(i, j) + \left(\lambda \mathbf{I}\,(\mathbf{i}, \mathbf{j}) + \beta \mathbf{H}\,(\mathbf{i}, \mathbf{j})^T \mathbf{H}\,(\mathbf{i}, \mathbf{j})\right)^{-1}$$

$$\left(\lambda \mathbf{u}\,(\mathbf{i}, \mathbf{j})_k + \mathbf{u}\,(\mathbf{i}, \mathbf{j})_0\,\beta \mathbf{H}\,(\mathbf{i}, \mathbf{j})^T\right) + 2\alpha \Delta t \left(D_{1_k}\,(i, j) + D_{2_k}\,(i, j)\right) - \lambda \left(\mathbf{u}_k\,(i, j) - \mathbf{u}_0\,(i, j)\right)$$

$$(20)$$

for $k = 0, \cdots, K$, where u_0 represents the discrete form $[M \times N]$ of the initial degraded image and the boundary conditions are as follows:

$$\mathbf{u}_k\,(0, j) = \mathbf{u}_k\,(1, j), \quad \mathbf{u}_k\,(M + 1, j) = \mathbf{u}_k\,(M, j)$$
$$\mathbf{u}_k\,(i, 0) = \mathbf{u}_k\,(i, 1), \quad \mathbf{u}_k\,(i, N + 1) = \mathbf{u}_0\,(i, N)$$

The explicit numerical approximation procedure by Eq. (3) is stable and consistent to the nonlinear anisotropic diffusion model by Eq. (8). The pseudocode for the proposed model is shown in Algorithm (1):

Algorithm 1 Pseudo code for the proposed model

Input: $\alpha > 0, \beta > 0, \lambda > 0, \epsilon > 0, \Delta t, k \leq n, u_0$;
Output: reconstructed image $\mathbf{u}_r = \mathbf{u}_{(k+1)}$;
1: $k = 0$
2: TVM

$$\mathbf{E}\,(\mathbf{u}) = \int_\Omega \left(\frac{\beta}{2} \|\mathbf{H}\mathbf{u} - \mathbf{u}_0\|_2^2 + \alpha \|\nabla \mathbf{u}\| + \frac{\lambda}{2} \|\mathbf{u} - \mathbf{u}_0\|_2^2\right) \partial \Omega$$

3: **while** $k \leq n$ **do**
4:     Compute

$$\mathbf{u}_{k+1} = \mathbf{u}_k + \left(\lambda \mathbf{I} + \beta H^T H\right)^{-1} \left(\beta H^T \mathbf{u}_0 + \lambda \mathbf{u}_k\right) + 2\alpha \Delta t \left(D_{1_k} + D_{2_k}\right) - \lambda \left(\mathbf{u}_k - \mathbf{u}_0\right)$$

    where α and β estimates by Eq. (3);
5:     Checking is stopping criteria if k attains given iteration n and stop;
6:     $k = k + 1$ and return to Step 4;
7:     Output $\mathbf{u}_r = \mathbf{u}_{(k+1)}$;
8: **end while**

```
//kernel function for cuVariationalKernel
__global__ void cuVariationalKernel(float *u,float *D_2,float *D_1,float *u_0, float
    *u_x,float *u_y,float *u_xx,float *u_yy,float *u_xy, const int nx, const int ny,
    float dt,float beta,float alpha, float eps, float lambda, const int n,float *bln,
    float *bld,float *h,float *h_tr){
//parameters and all input matrices provide from host(CPU)
unsigned int ix = threadIdx.x + blockIdx.x * blockDim.x;
unsigned int iy = threadIdx.y + blockIdx.y * blockDim.y;
unsigned int idx = iy*nx + ix; unsigned int left = idx − 1;
unsigned int right = idx + 1; //boundary adjustment;
if (ix == 0) left++; if (ix == nx−1) right−−; int k;//n iteration;
while(k<n){
if (ix < nx && iy < ny){
u_x[idx] =(u[right−1] − u[left+1])/2; u_y[idx] = (u[right+1]− u[left+1])/2;
u_xx[idx] = (u[right−1] − 2*u[idx] + u[left+1]);
u_yy[idx] =( u[right+1] − 2*u[idx] + u[left−1]);
u_xy[idx] =( u[right−1] + u[left+1] −u[right+1]− u[left−1]) / 4;
D_2[idx] = u_xx[idx]*u_y[idx]*u_y[idx]− 2*u_x[idx]*u_y[idx]*u_xy[idx]
+ u_yy[idx]*u_x[idx]*u_x[idx]; D_1[idx] = pow((u_x[idx]*u_x[idx]
+ u_y[idx]*u_y[idx]),3/2)+ psilon;
bln[idx]=((beta*h_tr[idx]*u_0[idx]+lambda*u[idx]));
bld[idx]=(lambda*Ind[idx]+(beta*h_tr[idx]*h[idx]));
u[idx] = u[idx] + (bln[idx]/bld[idx]) +2*alpha*dt*( D_2[idx]+D_1[idx])
− 2*lambda*(u[idx]−u_0[idx]);
__ syncthreads(); //thraed synchronization
k=k+1;
        }
    }//end loop
}//end kernel;
```

4   Implementation of CUDA-Based Parallel Algorithm

This section describes the design and parallel implementation of the proposed algorithm on the GPU in detail. Special GPU card is dedicated for scientific computing, like the NVIDIA Quadro $K420$ card which is utilized in this work to perform the entire experiment. Such a GPU card is composed of a number of 192 CUDA cores with 1 GB ECC memory, shared by all processor cores [12, 13]. Moreover, NVIDIA CUDA, a general-purpose parallel computing architecture with an advanced parallel programming model uses a set of instructions that powers the parallel computation in NVIDIA GPUs to solve various complex problems in a more efficient way than on a CPU. CUDA comes with a software environment that allows developers to utilize C or C++ as high-level programming languages to overcome the challenges for developing application software and its parallelism to control the increasing number of processor cores [14]. In this work, OpenGL and the CUDA parallel programming models are successfully utilized to design and implement the proposed algorithm. The proposed model is computationally high, and to design it in multiple kernels, it involves several operations with big matrices. In the heterogeneous platform, GPU device is used to accelerate the multiple kernel-related computations and the CPU is used as a host to perform other data manipulation and control the operations. Details of the parallel implementation are given next.

First, using input/output (I/O) transfers between host and device, we allocate memory on the device, copy the original data from the host to the device at the beginning of the algorithm, and then transfer the final results from the device to the host at the end [12, 13]. In the second step, computation of the parameter estimation for each input image is performed individually by invoking a kernel. It takes full advantage of shared memories of the GPU to gain high efficiency in the computation of big image matrices. Next, kernel functions are designed to perform the gradient estimation on each input image separately. Depending on the different sizes of the input images, we begin at 32 × 32 block grid and a 32 × 32 thread grid on the GPU to compute in all kernels, respectively, where each thread calculates a single pixel. The number of threads per block is set to 32 × 32, based on computing capabilities of NVIDIA Quadro K420 Card [12, 13]. Next, another kernel function is designed to compute the second-order derivative for data matrix. It processes 32 × 32 block and 32 × 32 thread. Then, the result is stored in a different device variable. Finally, we implement a special kernel function called as *"cuVariationalKernel"* in the proposed algorithm with the C++. Each kernel executes the pixel-wise calculation of the image matrix with the iterative routing. The execution time of the *"cuVariationalKernel"* is high because it is an iterative process.

5 Experimental Analysis

5.1 Performance Analysis

To analyze the performance of the proposed technique, test images have taken from the medical image database [15]. Test images are used for the performance evaluation, which is shown in Fig. 1. Figure 1 is a fetal ultrasound or a sonogram image. In real time, the ultrasound image is automatically induced by the random noise (speckle noise, Gaussian noise) which is approximated by some standard noise generation algorithms. To generate a noisy image, the amount of noise is added to the image by the Gaussian distribution with varying variance. In this paper, Gaussian noise has been added to each ultrasound image using standard deviation with zero means and reflected as an additive noise ultrasound image. The effects of adding noise to the test images are clearly shown in Fig. 2. The effects of Gaussian noise have significantly degraded the visibility of the ultrasound image and distorted the entire boundaries of the image objects. Furthermore, Gaussian noise is extended additively throughout whole regions of the ultrasound image.

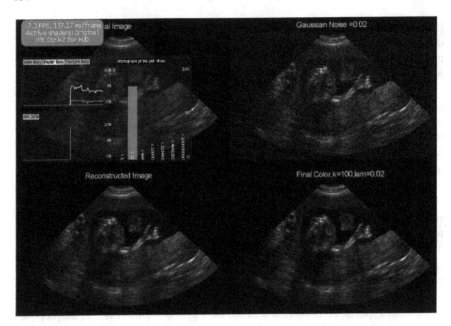

Fig. 1 Restoration results for ectopic pregnancy ultrasound images by the proposed model on the CUDA GPUs

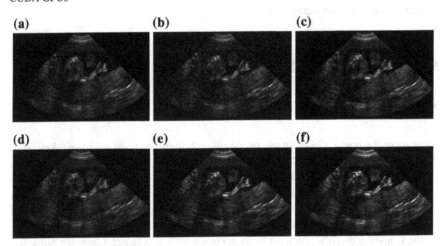

Fig. 2 Reconstruction results obtained by several restoration techniques: **a** original image, **b** noise version, **c** "variational PDE technique" [4], **d** "variational and combination model" [6], **e** "iterative decoupling variational algorithm" [8], and **f** the proposed method

5.2 Result Analysis

In our experiment, the proposed algorithm deals with a real recorded ultrasound medical images such as pericardial effusion sonogram of a patient [15]. In this case, the pericardial effusions occur when excess fluid accumulates in the pericardial space in between the parietal and visceral layers of the serous pericardium [15]. For simulation purpose, experimental images have been resized with noise from the source image maintaining the same original resolution. To evaluate the performance of the proposed algorithm on the GPU, different sizes of noisy images are tested such as 512×512, 1024×1024, 2048×2048, and 4096×4096 after resizing the original image size. The sequential and parallel execution of the proposed algorithm on the CPU and GPU can lead to different time resolutions over several image sizes. The reconstructed result of the proposed algorithm is shown in Fig. 1. The time consumption on GPU processing of the proposed method is given in Table 1 with the unit as a second. From the result, we can say that, to deal with the small data, the CPU is better than the GPU. The speedup of the proposed method is shown in Table 2. As illustrated in Table 2, while the proposed algorithm executes on the GPU in parallel, the execution accelerates (speedup) twenty orders of scale.

In this work, to test the quality of restoration, the proposed algorithm is compared with few state-of-the-art restoration algorithms. We select three restoration approaches such as the "variational PDE technique" [4], "variational and combination model" [6], and the "iterative decoupling variational algorithm" [8]. Moreover, our main goal is to explore the performance variation of the proposed algorithm with several image resolutions on Quadro-k20 GPU platform. On the other hand, we apply two metrics signal-to-noise ratio (**SNR**) and relative error (**ReErr**) on all resulting images obtained by the four algorithms separately [1, 2]. Table 1 summarizes the **SNR** and **ReErr** results achieved. Four noise sensitivities ($\sigma = 0.01, 0.05, 0.10, 0.20$) are used in every single experiment, and the achieved **SNR** and **ReErr** for a pair of pericardial effusions ultrasound images are listed in Table 1. By the assessment of Table 1, we can notice that for better denoising results the proposed algorithm generates the best values of **SNR** and **ReErr,** respectively. To justify a visual comparison of the restored images, we demonstrate the restored results with four noise levels by four methods for an ultrasound image as shown in Fig. 2. From the experimental results, the proposed technique achieves visually enhanced results than the other three approaches. The comparison results of four methods are shown and labeled with bold in

Table 1 Computational complexity of the proposed model on the CPU–GPU

Computational time on the CPU and GPU (iteration $k = 150$)				
Image	Resolution	CPU(s)	GPU (s)	Speedup [13, 14]
"Image" [1]	512×512	73.7413	3.29218	22.4
	1024×1024	135.157	6.08813	22.2
	2048×2048	205.168	9.20074	22.3
	4096×4096	281.736	12.4662	22.6

Table 2 Quantitative assessment of different restoration techniques

Quantitative results of the **SNR** and **ReErr** with four Gaussian noise

Image	Noise (σ)	Method [4]		Method [6]		Method [8]		Proposed	
		SNR	ReErr	SNR	ReErr	SNR	ReErr	SNR	ReErr
Image [2]	0.01	11.837	0.0133	11.819	0.0136	11.751	0.0129	11.901	0.0121
	0.05	9.8764	0.0151	9.8673	0.0154	9.6732	0.0143	9.8973	0.0118
	0.10	8.7653	0.0164	8.7318	0.0171	8.6185	0.0148	9.4091	0.0162
	0.20	8.0199	0.0198	8.0156	0.0202	8.1364	0.0198	9.1643	0.0183

Fig. 2 and Table 1. From Table 2, we notice that the proposed method is faster on the GPU than on the CPU. So, from Table 2 and Table 1, we establish that the proposed method performance is better on the GPU than on the CPU.

6 Conclusion

In this paper, we present a fast GPU-based denoising and deblurring approaches for the distorted medical ultrasound images. The scheme develops the technique of total variational model to remove noise and blur effect from medical images on the GPU. The proposed algorithm is derived from the second-order nonlinear partial anisotropic diffusion with the minimization of the total variational model. This model is discretized by applying the finite-difference-based numerical approximation technique. The efficiency of the method is tested with the ultrasound images embedded with additive Gaussian noise and blur. Experimental results show that the proposed technique achieves the best speedup on the GPU than the sequential execution on the CPU.

The authors declare that the test images used in this study are all public. The medical image database is available at [15]

References

1. Chan, T., Esedoglu, S. et al., Recent developments in total variation image restoration. Mathematical Models of Computer Vision. Springer, New York (2005)
2. Aubert, G., Kornprobst, P.: Mathematical Problems in Image Processing. Springer Science, New York (2006). ISBN 978-0387-32200-1
3. Barbu, T., Barbu, V.: PDE approach to image restoration problem with an observation on a meager domain. Nonlinear Anal. Real World Appl.,vol.13, no. 3, pp. 1206–15, 2012
4. Barbu, T., et al.: A novel variational PDE technique for image denoising, pp. 501–508. Springer, Heidelberg (2013)
5. Barbu, T., Favini, A.: Rigorous mathematical investigation of a nonlinear anisotropic diffusion-based image restoration model. Electron. J. Differ. Eqs. **129**, 1–9 (2014)
6. Vega, M., Mateos, J., et al.: Astronomical image restoration using variational methods and model combination. Statistical Methodology **9**, 19–31 (2012)
7. Xu, J., Feng, A., et al.: Image deblurring and denoising by an improved variational model. Int. J. Electron. Commun. (AUE) **70**(9), 1128–1133 (2016)
8. Wen, Y., Ng, M., Ching, W.: Iterative algorithm based on decoupling of deblurring and denoising for image restoration. SIAM J. Sci. Comput. **30**(5), 265574 (2008)
9. Liu, G., Huang, T., Liu, J.: High-order TVL1-based images restoration and spatially adapted regularization parameter selection. Comput. Math Appl. **67**, 201526 (2014)
10. Takeda, H., Farsiu, S., Milanfar, P.: Deblurring using regularized locally adaptive kernel regression. IEEE Trans. Image Process. **17**(4), 550–563 (2008)

11. Huang, J., Huang, T., et al.: Two soft-thresholding based iterative algorithms for image deblurring. Inf. Sci. **271**, 179–195 (2014)
12. Beck, A., Teboulle, M.: Fast gradient-based algorithms for constrained total variation image denoising and deblurring problems. IEEE Trans. Image Process. **18**(11), 2419–2434 (2009)
13. NVIDIA, CUDA: Compute Unified Device Architecture. http://docs.nvidia.com/cuda/# axzz3apxVEBRJ
14. Kandrot, E., Sanders, J.: CUDA by Example. Addison Wesley, Boston (2011)
15. http://www.mypacs.net

Line, Word, and Character Segmentation from Bangla Handwritten Text—A Precursor Toward Bangla HOCR

Payel Rakshit, Chayan Halder, Subhankar Ghosh and Kaushik Roy

Abstract The basic functionalities of optical character recognition (OCR) are to recognize and extract text to digitally editable text from document images. Apart from this, an OCR has other potentials in document image processing such as in automatic document sorter, writer identification/verification. In current situation, various commercially available OCR systems can be found mostly for Roman script. Development of an unconstrained offline handwritten character recognition system is one of the most challenging tasks for the research community. Things get more complicated when we consider Indic scripts like Bangla which contains more than 280 modified and compound characters along with isolated characters. For recognition of handwritten document, the most convenient way is to segment the text into characters or character parts. So line, word and character level segmentation plays a vital role in the development of such a system. In this paper, a scheme for tri-level segmentation (line, word, and character) is presented. Encouraging segmentation results are achieved on a set of 50 handwritten text documents.

Keywords OCR · Bangla handwritten character recognition · Line segmentation
Word segmentation · Character segmentation

P. Rakshit (✉) · C. Halder · S. Ghosh · K. Roy
Department of Computer Science, West Bengal State University,
Barasat, Kolkata 700126, West Bengal, India
e-mail: prmylife20@gmail.com

C. Halder
e-mail: chayan.halderz@gmail.com

S. Ghosh
e-mail: sgcs2005@gmail.com

K. Roy
e-mail: kaushik.mrg@gmail.com

© Springer Nature Singapore Pte Ltd. 2018   109
R. Chaki et al. (eds.), *Advanced Computing and Systems for Security*,
Advances in Intelligent Systems and Computing 666,
https://doi.org/10.1007/978-981-10-8180-4_7

1   Introduction

In recent era of digital evaluation, computer-aided document processing (reading/writing) is getting more importance in our day-to-day life. Here, optical character recognizer (OCR) will be utilitarian, if developed properly. Character recognition of printed text document has achieved a great success rate following huge interest by researchers. Commercially available OCR systems like fine reader by ABBYY is one of the prime examples [1]. Various offline handwritten optical character recognition (HOCR) strategies for non-Indic scripts such as English [2], Chinese [3], Japanese [4] are already proposed by different authors but only a few stray works are done on offline Bangla Handwritten character recognition [5–12]. The impediments behind are inconstant variations of human writing style and similarities of distinct character shapes, overlapping and touching of neighboring characters, spatial variation of characters when combined with other characters (modified complex and compound characters), etc. [6]. Hence, an efficient Bangla OCR system needs to be developed for recognition of handwritten text.

In Indian subcontinent, Bangla is the second most popular script after Devanagari [13]. Many research works are already investigated for handwritten character recognition of different Indic scripts. Sahlol et al. proposed an OCR algorithm for recognition of handwritten Arabic characters [14], a deep learning-based large-scale handwritten Devanagari characters recognition scheme was proposed by Acharya et al. [15], Kamble and Hegadi [16] came up with an idea for handwritten Marathi character recognition, whereas Varghese et al. [17] developed a novel recognition tri-stage scheme for recognition of handwritten Malayalam characters. Some works on Bangla isolated alphabets and numerals are also proposed by Rahman et al. [12], Sarkhel et al. [9], and Wen et al. [8]. Halder and Roy [10] suggested a method for Bangla Handwritten character segmentation from words. Das et al. [7] presented a system for handwritten Bangla basic and compound character recognition. A structural composition based Bangla compound characters recognition strategy was provided by Bag et al. [6].

In this proposed work, an attempt is made for development of a tri-level (line, word, and character) segmentation scheme without any normalization. The rest of the paper is organized as follows: Sect. 2 describes the overview of the study, methodology is presented in Sect. 3, whereas results are shown in Sect. 4 and finally, we concluded in Sect. 5.

2   Overview of the Study

A tri-level segmentation (line, word, and character) scheme which is an essential part of an OCR is proposed here in this work. In Fig. 1, a block level overview of an OCR system is depicted. The most enlightened area of this work is, during segmentation, no normalization is performed, i.e., an attempt is made toward developing a HOCR

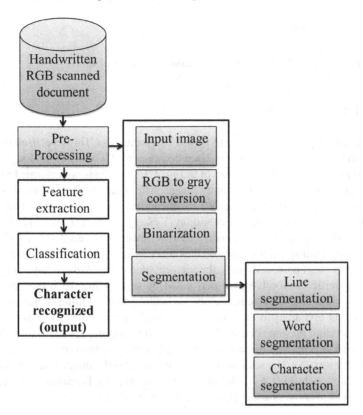

Fig. 1 Block diagram of the proposed HOCR system

without any noise removal and skew/slant correction. In the proposed model, upto preprocessing part is implemented marked as shaded regions in Fig. 1. The remaining stages of feature extraction and classification will be applied on the segmented characters obtained from the current work.

3 Methodology

In this stage, the raw scanned RGB document images are processed to be used for segmentation. Firstly, extraction of only handwritten text is done using a text extraction technique. In this technique, first of all horizontal and vertical boundary lines are detected using histogram localization. After that, these identified boundary lines and all other unnecessary printed texts are deleted to extract only handwritten text part applying minimum bounding box. The extracted grayscale images are then stored in tif format.

3.1   Segmentation

This part consists of a tri-level segmentation as shown in Fig. 1.

3.1.1   Line Segmentation

Line segmentation is the first level of tri-level segmentation. This level itself consists of different sub-levels such as connected component (CC) analysis, Hough transform, filling, and smoothing as shown in Fig. 2. CC analysis is performed to mark word components (WCs) in the binary image. After that, the average word component height (WCh_{avg}) and average WC width and total number of WCs (tot_{wc}) are calculated. Here, we have used a condition ($TH_{min} \leq WC \leq TH_{max}$) to eliminate too large and too small WCs depending on two threshold values TH_{min}, TH_{max}.

Now Hough transform is applied on the binary image to estimate the potential text lines and their start points. To make the segmentation easier, bidirectional horizontal filling and vertical smoothing are performed by segmenting the image into Seg_n (here, $Seg_n = 8$ is considered empirically) vertical segments. Filling operation is used to fill the non-text areas in the image, and smoothing is applied immediately after filling, depending on some certain threshold to fill the small gaps in the filled image. Each transition is marked by two points, namely start transition ($trans_{start}$) and end transition ($trans_{end}$). The total number of transitions in the image is denoted as (t). Average line height (WCh_{avg}) is calculated using Eq. (1). Equation (2) represents calculation of average line gap (LG_{avg}) between two lines.

$$WCh_{avg} = \frac{\sum_{i=0}^{tot_{wc}-1}(WC_{height})_i}{tot_{wc}} \qquad (1)$$

where WC_{height} represents height of each word component.

Fig. 2 Steps of line segmentation

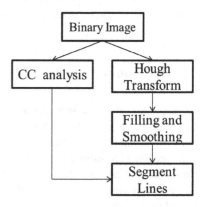

$$LG_{avg} = \frac{\sum_{i=0}^{t-1}(trans_{end} - trans_{start})_i}{t} \tag{2}$$

Algorithm 1: Filling of gray image with smoothing

Input : Handwritten gray text image I_gray.
Output: Space filled image I_gray$_{smooth\_filled}$.

1   **for** $s \leftarrow 0$ **to** Seg_n **do**
     // Seg_n is the number empirically taken to divide the text for filling
2     $Spoint_{seg} \leftarrow 0$;
3     $Epoint_{seg} \leftarrow 0$;
4   **for** $i \leftarrow 0$ **to** $height$ **do**
       // Left to right filling
5   diff $\leftarrow$ width/Seg_n;
6       $V \leftarrow 0$;
7       $V \leftarrow V + (s + diff)$;
8       $Spoint_{seg} \leftarrow Epoint_{seg}$;
9       $Epoint_{seg} \leftarrow V$;
10      $Spoint_{seg} \leftarrow$ **for** $j \leftarrow Spoint_{seg}$ **to** $Epoint_{seg}$ **do**
11   **if** $I\_gray(i, j) \neq Object\,Pixel$ **then**
12   I_gray$_{filled}$(i,j) $\leftarrow$ Fill_color;
13   **end**
14   **end**
       // Right to left filling
15   **for** $j \leftarrow Epoint_{seg}$ **to** $Spoint_{seg}$ **do**
16   **if** $I\_gray(i, j) \neq Object\,Pixel$ **then**
17   I_gray$_{filled}$(i,j) $\leftarrow$ Fill_color;
18   **end**
19   **end**
20   **end**
21 **end**
    // smoothing
22 **for** $i \leftarrow 0$ **to** $width$ **do**
23   **for** $j \leftarrow 0$ **to** $height$ **do**
24   **for** $ss \leftarrow trans_{start}$ **to** $trans_{end}$ **do**
25        $counter \leftarrow counter + 1$;
26   **end**
27       $smt_{thresh} \leftarrow WCh_{avg} * 20\%$;
28   **if** $count \geq smt_{thresh}$ **then**
29   I_gray$_{smooth\_filled}$(i,j) $\leftarrow$ Fill_color;
30   **end**
31   **end**
32 **end**

The detected lines are segmented by running a separator through the interline gap between two lines of the smoothed image. Another step is employed to detect and

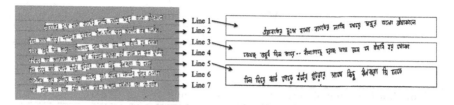

Fig. 3 Lines in a segmented image

separate the text lines that Hough transform failed to detect. The segmentation of the lines is started from their start points, and the separator moves in straight line in forward direction (from left to right) through the filled non-text area between two text lines. At that time, if the separator touches a text pixel (black) or text area (white), a decision making is performed at that point. The separator checks in upward and downward direction through which it can move forward (Fig. 3). If both of the directions are closed, the separator cuts through the text area. After completing this operation, we have a basic set of lines. Now each segmented line height is checked again. If the line height of any segmented line is greater than the threshold LH_{th}, the line is checked again whether there exist multiple lines. Equation (3) is used to calculate LH_{th}. This situation occurs because sometimes Hough transform fails to detect all the lines, and multiple lines are treated as a single line. To segment these lines again, same segmentation procedure is repeated. Finally, we have the segmented set of all lines. A sample segmented image is shown in Fig. 3.

$$LH_{th} = (2 \times (WCh_{avg} + LG_{avg})) \times 0.8 \tag{3}$$

3.1.2 Word Segmentation

In the proposed scheme, word segmentation is performed based on CC analysis. Before applying CC analysis, morphological erosion, and dilation are performed depending on erosion and dilation threshold. In the previous section (line segmentation), CC analysis is already discussed. Here, each WC of a line is treated as a word. As in Bangla script, "matra" or "shirorekha" is mostly used to connect the characters of a word; therefore, words can be recognized easily by identifying the CCs. The CCs of a line are shown using bounding boxes which represent the words in Fig. 4.

3.1.3 Character Segmentation

Zone segmentation is required to segment the characters properly as Bangla words contain character parts (modifiers) in upper and lower zones along with middle zone. In this scheme, busy zone, headline, and baseline information are used to separate

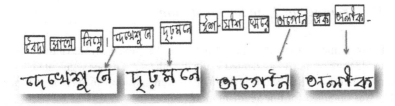

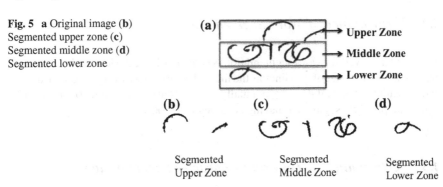

Fig. 4 Segmented words from a line

Fig. 5 a Original image (**b**)
Segmented upper zone (**c**)
Segmented middle zone (**d**)
Segmented lower zone

three zones. After segmenting the three zones of a word, character segmentation of the base characters of the middle zone is necessary for proper segmentation. Middle zone segmentation is performed by using the interspace between two characters and vertical projection profile method. After identifying the segmentation points, the characters of the middle part are vertically segmented using those segmentation lines. The details about this character segmentation technique can be found in [10] (Fig. 5).

4   Results

The current study of three-level segmentation is employed on a dataset of 50 Bangla unconstrained handwritten text documents from same number of individuals. No restriction is posed on the type of writing instruments used. The dataset contains wide variation of distinct writing style because the writers are of different age, gender, educational qualification. The collected datasets are scanned and stored in 300 dpi RGB mode. Currently, no ground truth data is available, so we have calculated segmentation accuracy manually.

4.1   Performance of Line Segmentation

Table 1 shows line segmentation performance of the current work. Here out of total 482 lines 436 are properly segmented while the others are under segmented. No case of over-segmentation has occurred during the segmentation. An average accuracy of 90.46% was obtained for line segmentation.

4.2   Performance of Word Segmentation

The result achieved for word segmentation is shown in Table 2. Here we have considered total 200 lines out of 436 lines as the manual checking was very time-consuming and difficult to check all the lines manually. Out of 1640 words from 200 lines, 1477 words are properly segmented having accuracy of 90.06%.

4.3   Performance of Zone and Character Segmentation

Here we have considered a set of 500 words for manual checking. The detail result of zone segmentation is shown in Table 3. The accuracies achieved for upper zone, middle zone, and lower zone are 66.56%, 98.80%, and 75.97%, respectively. In Table 4, one can see that out of 3015 characters, 1524 are properly segmented obtaining an average accuracy of 50.55%.

4.4   Error Analysis

In this proposed method, it is found that the line segmentation and word segmentation accuracies are encouraging enough to be used in Bangla OCR. In case of word segmentation, erroneous situations are occurred when word components of a same word

Table 1 Line segmentation result of all the documents

Documents	Total lines	Segmented lines	Accuracy (%)
50	482	436	**90.46**

Table 2 Word segmentation result of segmented lines

Lines	Total words	Segmented words	Accuracy (%)
200	1640	1477	**90.06**

Table 3 Zone segmentation result of segmented words

Words	Total zones of characters			Segmented zones of characters			Accuracy (%)			Average accuracy (%)
	Upper	Middle	Lower	Upper	Middle	Lower	Upper	Middle	Lower	
500	308	500	195	205	494	122	66.56	98.80	62.56	**75.97**

Table 4 Character segmentation result of segmented words

Words	Total characters	Segmented characters	Accuracy (%)
500	3015	1524	**50.55**

(a)

(b)

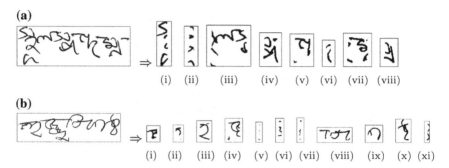

Fig. 6 a–b Some different word examples which causes the errors of character segmentation

are written with significant distance between them which leads to over-segmentation, similarly when two words are touching each other then under segmentation occurs. The character segmentation accuracy is not as high as the existing results due to the fact that segmentation of handwritten text has various challenges that are discussed earlier in Sect. 1. Apart from those, it should also be noted that in this proposed method any kind of correcting measures like skew and slant corrections are not considered. In Fig. 6, erroneous character segmentation of two sample words is presented. In Fig. 6a, the word is multi-oriented creating a multi-level skew with irregularly placed characters at different zones, along with this the characters are also touching which leads to improper character segmentation result. The similar multi-level skew can also be observed in Fig. 6b.

5 Conclusion

In this paper, line, word, and character segmentation of unconstrained handwritten Bangla text, document is presented. The segmentation is one of the most essential and preliminary tasks for many document image processing works. The task is more challenging when it needs to be developed on unconstrained handwriting. The proposed approach of line segmentation is a hybrid approach to improve line segmentation accuracy compared to existing methods. A simple yet effective word segmentation procedure is presented based on connected component analysis. Character segmentation is performed on a relatively large number of words set using an existing zone and character segmentation method. The prime factor of low accuracy for the proposed work is due to existence of skew, slant, touching/overlapping components, and lack of connections between two consecutive characters of a word. Here, no attempt is made to correct and normalize this natural randomness of writing to make the method simple, so that it becomes easier to be incorporated in most of the OCR systems. The main drawback of this technique lies in word and character segmentation area where it fails to reduce over-segmentation.

In future, to deal with these touching and overlapping writing we will introduce combination of different approaches which will improve the segmentation accuracy. Further, these segmented characters will be used for feature extraction and classification to observe character recognition results in recent future. We have planned to use chain code-based features, histogram gradient-based features, and local binary pattern (LBP) for feature extraction. Different renowned classifiers like MLP, SVM will be tested in future along with combination of classifiers.

Acknowledgements Two of the authors, Ms. Payel Rakshit and Mr. Chayan Halder, are thankful to Department of Science and Technology (DST) for their support as INSPIRE fellowship.

References

1. https://www.abbyy.com/ . Last accessed 07 Dec 2016
2. Rakshit, S., Basu, S.: Recognition of handwritten Roman script using Tesseract open source OCR engine. In: National Conference on (NAQC), pp. 141–145 (2008)
3. Tsukumo, J., Tanaka, H.: Classification of handprinted Chinese characters using nonlinear normalization methods. In: 9th International Conference on Pattern Recognition, pp. 168–171 (1988)
4. Yamada, H., Yamamoto, K., Saito, T.: A non-linear normalization method for handprinted Kanji character recognition line density equalization. Pattern Recognit. **23**, 1023–1029 (1990)
5. Bhunia, A.K., Das, A., Roy, P.P., Pal, U.: A comparative study of features for handwritten Bangla text recognition. In: 13th International Conference on Document Analysis and Recognition (ICDAR), pp. 636–640 (2015)
6. Bag, S., Harit, G., Bhowmick, P.: Recognition of Bangla compound characters using structural decomposition. Pattern Recognit. **47**, 1187–1201 (2013)
7. Das, N., Das, B., Sarkar, R., Basu, S., Kundu, M., Nasipuri, M.: Handwritten Bangla basic and compound character recognition using MLP and SVM classifier. J. Comput. **2**, 109–115 (2010)
8. Wen, Y., Lu, Y., Shi, P.F.: Handwritten Bangla numeral recognition system and its applicaiton to postal automation. Pattern Recognit. **40**, 99–107 (2007)
9. Sarkhel, R., Das, N., Saha, A.K., Nasipuri, M.: A multi-objective approach towards cost effective isolated handwritten Bangla character and digit recognition. Pattern Recognit. **58**, 172–189 (2016)
10. Halder, C., Roy, K.: Word & character segmentation for Bangla handwriting analysis & recognition. In: 3rd National Conference on Computer Vision, Pattern Recognition, Image Processing and Graphics, pp. 243–246 (2011)
11. Maitra, D.S., Bhattacharya, U., Parui, S.K.: CNN based common approach to handwritten character recognition of multiple scripts. In: 13th International Conference on Document Analysis and Recognition (ICDAR), pp. 1021–1025 (2015)
12. Rahman, M.M., Akhand, M.A.H., Islam, S., Shill, P.C., Rahman, M.M.H.: Bangla handwritten character recognition using convolutional neural network. Int. J. Image Graph. Signal Process. (IJIGSP) **7**, 42–49 (2015)

13. Halder, C., Obaidullah, S.M., Roy, K.: Effect of writer information on Bangla handwritten character recognition. In: 5th National Conference on Computer Vision, Pattern Recognition, Image Processing and Graphics (NCVPRIPG), pp. 1–4 (2015)
14. Sahlol, A.T., Suen, C.Y., Elbasyouni, M.R., Sallam, A.A.: A proposed OCR algorithm for the recognition of handwritten Arabic characters. J. Pattern Recognit. Intell. Syst. 2, 8–22 (2014)
15. Acharya, S., Pant, A.K., Gyawali, P.K.: Deep learning based large scale handwritten Devanagari character recognition. In: 9th International Conference on Software, Knowledge, Information Management and Applications (SKIMA), pp. 1–6 (2015)
16. Kamble, M., Hegadi, S.: Handwritten Marathi character recognition using R-HOG feature. In: International Conference on Advanced Computing Technologies and Applications (ICACTA), Procedia Computer Science, vol. 45, pp. 266–274 (2015)
17. Varghese, K.S., Jamesa, A., Chandran, S.: A novel tri-stage recognition scheme for handwritten Malayalam character recognition. In: International Conference on Emerging Trends in Engineering, Science and Technology (ICETEST), Pattern Recognition, vol. 24, pp. 1333–1340 (2016)
18. Htike, T., Thein, Y.: Handwritten character recognition using competitive neural trees. Int. J. Eng. Technol. 5, 352 (2013)

Heterogeneous Face Matching Using ZigZag Pattern of Local Extremum Logarithm Difference: ZZPLELD

Hiranmoy Roy and Debotosh Bhattacharjee

Abstract A novel methodology for matching of heterogeneous faces, such as sketch-photo and near-infrared (NIR)-visible (VIS) images is proposed here. For heterogeneous face recognition, more emphasis is given to the edge features, which are invariant in different modality images. Since edges are sensitive to illuminations, we present an illumination-invariant image representation called local extremum logarithm difference (*LELD*). *LELD* provides illumination-invariant edge features in coarse level. Therefore, a local zigzag binary pattern *LZZBP* is presented to capture the local variation of *LELD*, and we call it a zigzag pattern of local extremum logarithm difference (*ZZPLELD*). We tested the proposed methodology on different sketch-photo and NIR-VIS benchmark databases. In the case of viewed sketches, the rank-1 recognition accuracy of 96.35% is achieved on CUFSF database. In the case of NIR-VIS matching, the rank-1 accuracy of 99.39% is achieved and which is superior to other state-of-the-art methods. We also tested *ZZPLELD* on illumination variation Extended Yale B database, and rank-1 recognition accuracy of 94.51% is achieved.

Keywords Illumination-reflectance model · Modality-invariant feature
Illumination-invariant feature · Local extremum logarithm difference
Local zigzag binary pattern · Heterogeneous face recognition

H. Roy (✉)
Department of Information Technology, RCC Institute of Information Technology,
Canal South Road, Beliaghata, Kolkata 700015, India
e-mail: hiranmoy.roy@rcciit.org

D. Bhattacharjee
Department of Computer Science and Engineering, Jadavpur University,
Kolkata 700032, India
e-mail: debotoshb@hotmail.com

© Springer Nature Singapore Pte Ltd. 2018
R. Chaki et al. (eds.), *Advanced Computing and Systems for Security*,
Advances in Intelligent Systems and Computing 666,
https://doi.org/10.1007/978-981-10-8180-4_8

121

1   Introduction

Biometric authentication is becoming the most essential and ubiquitous component of any modern systems, such as mobile phone, smart TV, computer. In different forensic applications different unique biometric features are used, such as fingerprints, faces, retinas, DNA samples, ears. Among these features, faces are the most easily available and easily recognizable feature. A single face biometric consists of a bundle of unique biometric features like eyes, nose. The valuable part of any face biometrics is that the recognition or authenticate can be done without any expertise. However, by the naked eye without expert, it is quite impossible to authenticate depending on fingerprints, DNA samples, retinas, and ears. Therefore, in recent years, a lot of works have been done on face biometric authentication based on real-life applications. Different real-life applications need faces accumulated in different environments. Near-infrared cameras are used to capture faces at night for illumination-invariant face recognition [1]. Thermal-infrared (TIR) cameras are used to capture body heat for liveness detection. Sometimes, it may happen that there are no available fingerprints, no available DNA samples, and devices have captured poor quality images. In those situations, face sketches generated by interviewing the eye witness are the only solution. Therefore, various scenarios and necessities create different modalities of faces. It becomes difficult to use conventional face recognition systems in those situations. Therefore, an interesting and challenging field of face biometric recognition has emerged for forensics, called heterogeneous face recognition (HFR) [2].

The problem of heterogeneous face recognition has received increasing attention in recent years. Up to now, many different techniques have been proposed in the literature to solve the problem. We can easily classify these solutions into three broad categories: image synthesis-based methods, common subspace learning-based methods, and modality-invariant feature representation-based methods.

- **Image synthesis**: In this category, a pseudo-face or pseudo-sketch is generated using synthesis techniques to transform one modality image into another modality and then some classification technique is used. The pioneering work of Tang and Wang [3], where they introduced an eigen transformation-based sketch-photo synthesis method. The same mechanism was also used by Chen et al. [4] for NIR-VIS synthesis. Gao et al. [5] proposed an embedded hidden markov model and a selective ensemble strategy to synthesize sketches from photos. Wang and Tang [6] again proposed a patch-based Markov random field (MRF) model for the sketch-photo synthesis. Li et al. [7] used the same MRF model for TIR-VIS synthesis. Gao et al. [8] proposed a sparse representation-based pseudo-sketch or pseudo-photo synthesis. Another sparse feature selection (SFS) and support vector regression for synthesis were proposed by Wang et al. [9]. Wang et al. [10] proposed a transductive learning-based face sketch-photo synthesis (TFSPS) framework. Recently, Peng et al. [11] proposed a multiple representation-based face sketch-photo synthesis. In this category, more emphasis is given on synthesis, which is a time-consuming and "task-specific."

- **Common subspace learning**: In this category, different modality face images are projected into a subspace for learning. Lin and Tang [12] introduced a common discriminant feature extraction (CDFE) for face sketch-photo recognition. Yi et al. [13] proposed a canonical correlation analysis-based regression method for NIR-VIS face images. A coupled spectral regression (CSR)-based learning for NIR-VIS face images was proposed by Lei and Li [14]. A partial least square (PLS)-based subspace learning method was proposed by Sharma and Jacobs [15]. Mignon and Jurie [16] proposed a cross modal metric learning (CMML) for heterogeneous face matching. Lei et al. [17] proposed a coupled discriminant analysis for HFR. A multi-view discriminant analysis (MvDA) technique for single discriminant common space generation was proposed by Kan et al. [18]. In this category, projection of images is performed, which generates some loss of information and at the same time reduces the accuracy.
- **Modality-invariant feature representation**: In this category, images of different modalities are represented using some modality-invariant feature representation. Liao et al. [19] used difference of Gaussian (DoG) filter and multi-block local binary pattern (MB-LBP) features for both NIR and VIS face images. Klare et al. [20] employed the scale-invariant feature transform (SIFT) and multi-scale local binary pattern (MLBP) features for forensic sketch recognition. A coupled information-theoretic encoding (CITE) feature was proposed by Zhang et al. [21]. Bhatt et al. [22] used a multi-scale circular Weber's local descriptor (MCWLD) for semi-forensic sketch matching. Klare and Jain [23] proposed a kernel prototype random subspace (KP-RS) on MLBP features. Zhu et al. [24] used a Log-DoG filter-based LBP and a histogram of oriented gradient (HOG) features with transductive learning (THFM) for NIR-VIS face images. Gong et al. [25] combined histogram of gradients (HOG) and multi-scale local binary pattern (MLBP) with canonical correlation analysis (MCCA). Roy and Bhattacharjee [26] proposed a geometric edge–texture-based feature with hybrid multiple fuzzy classifier for HFR. Roy and Bhattacharjee [27] again proposed an illumination invariant local gravity face (LG-face) for HFR. A local gradient checksum (LGCS) feature for face sketch-photo matching was proposed by Roy and Bhattacharjee [28]. Another local gradient fuzzy pattern (LGFP) based on restricted equivalent function for face sketch-photo recognition was again proposed by Roy and Bhattacharjee [29]. Recently, a graphical representation-based HFR (G-HFR) was proposed by Peng et al. [30]. In this category, local handcrafted features are directly used, which means no loss of local information and algorithms are more time saving than other two categories. One and only problem in this category is to recognize or search features, which are either common to different modalities or invariant in different modalities.

Modality-invariant feature representation methods are neither time-consuming and task-specific synthesis, nor common subspace-based learning but able to consider the local spatial features. Motivated by the advantages of modality-invariant feature representation, in this paper, we propose a modality-invariant feature representation for different modality images. The goal of the proposed method is to recognize the facial features, which are invariant in different modalities. From our visual inspection, we conclude that edges are the most important modality-invariant feature. Psychological studies also say that we can recognize a face from its edges only [31]. It is easy to understand that facial components in a face have maximum edges and they belong to high-frequency components of an image. Another feature, i.e., texture information is also important for face matching. Since edges and textures in a face image are sensitive to illumination variations, an illumination-invariant domain with the capability of capturing high-frequency information is necessary.

The artist gives more attention toward edges and texture information at the time of drawing a sketch. In the case of NIR images, the high-frequency information is captured. Therefore, selection of edge and texture features for modality-invariant representation is correct in the sense. We propose an illumination-invariant image representation called local extremum logarithm difference (*LELD*), which is a modification of the work explained in [32]. *LELD* gives only high-frequency image representation at a coarse level. A local micro-level feature representation is also important to capture local texture information. Motivated by the superior output results of the local binary pattern (LBP) [33] and LBP-like features in face and texture recognition, we propose one novel local zigzag binary pattern (*LZZBP*). *LZZBP* measures the binary relation between pixels, which are in a zigzag position in a local square window. *LZZBP* captures more edge and texture patterns than LBP. Finally, the combination of *LELD* and *LZZBP* gives the proposed modality-invariant feature representation for HFR and we call it a zigzag pattern of local extremum logarithm difference (*ZZPLELD*). Experimental results on different HFR databases show the excellent performance of the proposed methodology.
The major contributions are:

1. *LELD* is proposed for capturing illumination-invariant image representation.
2. *LZZBP* is developed to capture the relation between pixels in a zigzag position of a square window.
3. *ZZPLELD* is developed to capture the local texture and edge patterns of the modality-invariant key facial features.

This paper is organized as follows: in Sect. 2, the proposed *ZZPLELD* is described in detail. Experimental results and comparisons are presented in Sect. 3, and finally, the paper concludes with Sect. 4.

2 Proposed Work

In this section, we introduce the way we extract the modality-invariant features for HFR. We start with a detailed idea about illumination-invariant *ELLD*-based image representation. Finally, we conclude with a detail description of the proposed *ZZPLELD* feature.

2.1 Local Extremum Logarithm Difference Based Image Representation

In any face recognition system, one of the main problems is the presence of illu-mination variations. Due to the presence of illumination variations, the intra-class variation between faces also increases heavily. At the same time, we consider edges as our modality-invariant feature and edges are also sensitive toward illumination. Therefore, we need an illumination-invariant image representation for extracting bet-ter edge information. According to the illumination-reflectance model (IRM) [34, 35], a gray face image $I(x, y)$ at each point (x, y) is expressed as the product of the reflectance component $R(x, y)$ and the illumination component $L(x, y)$, as shown in Eq. 1

$$I(x, y) = R(x, y) \times L(x, y) \tag{1}$$

Here, the R component consists of information about key facial points, and edges, whereas the L component represents only the amount of light falling on the face. Now, after the elimination of the L component from a face image, the R compo-nent is still able to represent the key facial features and edges, which are the most important information for our modality-invariant feature representation. Moreover, the L component corresponds to the low-frequency part of an image, whereas the R component corresponds to the high-frequency part. One widely accepted assumption in the literature [27, 36] is that L remains approximately constant over a local 3×3 neighborhood.

In literature, a wide range of approaches have been proposed to reduce the illu-mination effect. In those methods, mainly two different mathematical operations are used: division and subtraction. Methods like [36, 37] are used division operation and methods like [32, 41] are used subtraction operation. In the case of subtrac-tion operation, at first, the image is converted to the logarithmic domain to convert the multiplicative IRM into an additive one, as shown in Eq. 2. Since the division operation is "ill-posed and not robust in numerical calculations" [32] due to the prob-lem of divided by zero, it is better to apply subtraction operation to eliminate the L component.

$$I_{log}(x, y) = log\,(R(x, y) \times L(x, y))$$
$$\Rightarrow I_{log}(x, y) = log\,(R(x, y)) + log\,(L(x, y)) \tag{2}$$

Lai et al. [32] proposed a multi-scale logarithm difference edgemaps (MSLDE), which used logarithmic domain and subtraction operation to compensate the illu-mination effect. They calculated a local logarithm difference using the following equation:

$$MSLDE = \sum_{(\hat{x}, \hat{y}) \in \aleph_{(x,y)}} \left(I_{log}(x, y) - I_{log}(\hat{x}, \hat{y}) \right), \forall x, \forall y \tag{3}$$

where $\aleph_{(x,y)}$ is a square neighborhood surrounding a center pixel at (x, y). For any single neighbor pixel and its center pixel, the logarithm difference is as follows:

$$MSLDE_1 = I_{log}(x, y) - I_{log}(\hat{x}, \hat{y})$$
$$\Rightarrow MSLDE_1 = (log\,(R(x, y)) + log\,(L(x, y))) - (log\,(R(\hat{x}, \hat{y})) + log\,(L(\hat{x}, \hat{y})))$$
$$\Rightarrow MSLDE_1 = (log\,(R(x, y)) - log\,(R(\hat{x}, \hat{y}))) + (log\,(L(x, y)) - log\,(L(\hat{x}, \hat{y}))) \qquad (4)$$

Since according to the assumption that, L component is almost constant in a small neighborhood. Therefore, Eq. 4 becomes:

$$MLDE_1 = (log\,(R(x, y)) - log\,(R(\hat{x}, \hat{y}))) \qquad (5)$$

Now, only R component is present in MSLDE. Therefore, MSLDE is an illumination-invariant method. In this method, authors considered a long range of square neighborhood from 3×3 to 13×13 and all the logarithm difference values are added together. Now, the question is does the L component be constant for such a long neighborhood, i.e., 13×13. We know that R component belongs to high-frequency. Similarly, edge and noise also belong to high-frequency. There is no doubt that MSLDE increases the edge information, which is very important in face recognition, by adding all logarithm differences. However, it is also increasing the noise, which causes degradation in true edge detection. The effect of noise is clearly visible in Fig. 1d, where the proposed MSLDE provides too much false edge information and which are nothing but noise. To solve both the problems, i.e., large neighborhood size and presence of noise, we consider the maximum and minimum logarithm difference in a 3×3 neighborhood and we call it local extremum logarithm difference (*LELD*). The maximum and minimum logarithm differences are calculated as follows:

$$LELD_{max} = max_{(\hat{x}, \hat{y}) \in \aleph_{(x,y)}} \left\{ I_{log}(x, y) - I_{log}(\hat{x}, \hat{y}) \right\}, \forall x, \forall y \qquad (6)$$
$$LELD_{min} = min_{(\hat{x}, \hat{y}) \in \aleph_{(x,y)}} \left\{ I_{log}(x, y) - I_{log}(\hat{x}, \hat{y}) \right\}, \forall x, \forall y \qquad (7)$$

where $\aleph_{(x,y)}$ is a 3×3 square neighborhood surrounding a center pixel at (x, y). Since we are considering the smallest square neighborhood, i.e., 3×3, the assumption that L is constant in the neighborhood is theoretically true. Again, we are avoiding total sum of differences; therefore, presence of noise is reduced. The results of proposed *LELD* are shown in Fig. 1e–h, and it gives better edge detection result than MSLDE. Finally, we have two different *LELD* methods to convert the different modality face images into an illumination-invariant domain, where important facial key values and edges are almost intact.

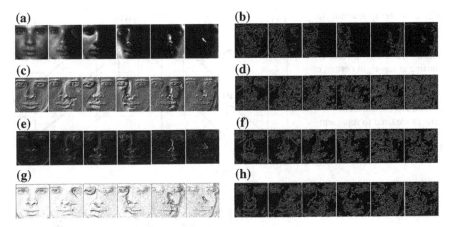

Fig. 1 **a** The original images with illumination variations, **b** The canny edge images for Fig. 1a images, **c** The corresponding MSLDE images of Fig. 1a images as proposed in [32], **d** The canny edge images of MSLDE images, **e** The proposed *LELD* images of Fig. 1a images with maximum difference, **f** The canny edge images of Fig. 1e images, **g** The proposed *LELD* images of Fig. 1a images with minimum difference, **h** The canny edge images of Fig. 1g images

2.2 *Local Zigzag Binary Pattern for* ZZPLELD *generation*

Local binary pattern (LBP) [33] has been used successfully in many fields of image processing and pattern classification problems. It is capable to represent local features in microstructures. LBP implements the binary relation of each and every neighboring pixels with respect to center pixel, i.e., if neighboring pixel is greater than or equal to center pixel, then binary value "1" otherwise "0." Although LBP captures the binary relations between surrounding neighboring pixels with the center pixel, it is not able to capture edge information properly, mainly in diagonal direction. Since our modality-invariant feature is edge related, a local pattern having good edge capturing capability is important. Inspired from the zigzag scanning pattern used for MPEG data compression in discrete cosine transform (DCT) domain, we developed a zigzag binary pattern for the pixels in a zigzag position of a square mask. Figure 2 shows the positions of 3×3 and 5×5 zigzag scanning used in our experiment. We use a left to right zigzag scanning (considering top-left pixel as the starting point) and right to left zigzag scanning (considering top-right pixel as the starting point) in each square mask. Figure 2a shows the 3×3 left zigzag scanning, Fig. 2b shows the 3×3 right zigzag scanning, and Fig. 2c shows the 5×5 left zigzag scanning. Again, we consider only 8 bits binary patterns to make the histogram feature vector length up to 256 bins. In case of 5×5 image pixels, we can have a total of 24 bits binary string. To make it with in a small range, we divide it into three 8 bits binary string. Different arrows in Fig. 2c show different sequences of binary string.

Let, the image pixels in a square window are collected in a zigzag pattern, as shown in Fig. 2a. Then, the pixels are stored in a linear array $Z_P = \{g_1, g_2, g_3, \ldots, g_P\}$,

Fig. 2 a Zigzag scanning in a 3 × 3 window starting at top-left corner, **b** Zigzag scanning in a 3 × 3 window starting at top-right corner, **c** Zigzag scanning in a 5 × 5 window starting at top left corner. Here 3 different arrows are used to represent 3 different sequences of 8 bits binary strings

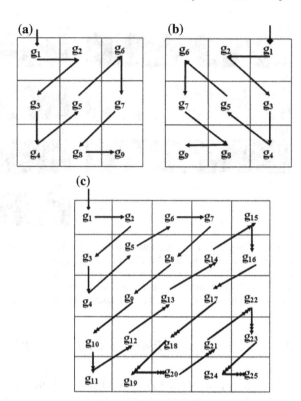

where P is the number of pixels in a square window. For 3 × 3 window P is 9 and for 5 × 5 window is 25. Then, we calculate the binary relation between those consecutive pixels according to the following equation (for $P = 8$):

$$LZZBP = \sum_{i=1}^{P-1} 2^{(i-1)} \times f\left(g_{i+1} - g_i\right)$$

$$f(a) = \begin{cases} 1, & if \quad a \geq 0 \\ 0, & otherwise \end{cases} \tag{8}$$

In case of 5 × 5 window, the 24 bits binary string is first broken into 3 parts of 8 bits string and then converted to three separate patterns.

The proposed zigzag binary pattern gives better edge preserving texture information. The different LBP images and *LZZBP* images in Fig. 3 show the edge preserving properties of *LZZBP*. The key facial features are preserved far better than normal LBP in *LZZBP*.

Finally, the proposed *LZZBP* is applied on *LELD* image representation to measure the local patterns of *LELD* image and which give our proposed *ZZPLELD*. For two different extremum *LELD*, i.e., maximum and minimum differences, we get 4

(a)   **(b)**

Fig. 3 **a** Sample image and its corresponding LBP and *LZZBP* images. The 2nd image is the 3×3 LBP image and rest 3rd -4th are 3×3 *LZZBP* images. **b** The 2nd image is the 5×5 LBP image and rest 3rd -8th are 5×5 *LZZBP* images

different *ZZPLELD* results after applying 3×3 *LZZBP* and 12 different *ZZPLELD* results after applying 5×5 *LZZBP*. Therefore, altogether 16 different *ZZPLELD* images. The image features are represented in the form of histogram bins.

2.3 Similarity Measure

The whole *ZZPLELD* image is divided into a set of non-overlapping square blocks with dimension $w_b \times w_b$. Then, the histogram of each square block is measured. Finally, all the histograms measured from all the blocks of all the different (16) *ZZPLELD* images are concatenated to obtain the final face feature vector. Here, we use the nearest neighbor (NN) classifier with histogram intersection distance measure. Therefore, for a query image (I_q) with a concatenated histogram $H_q^{i,j}$ and a gallery image (I_G) with a concatenated histogram $H_G^{i,j}$, the similarity measure for a particular level of *ZZPLELD* is given as follows:

$$S^k(I_q, I_G) = - \sum_{i,j} min\left(H_q^{i,j}, H_G^{i,j} \right) \qquad (9)$$

where $S^k(I_q, I_G)$ is the similarity score of kth *ZZPLELD* level of both query and gallery image; (i, j) is the jth bin of ith block. Square block selection is another essential task. In this paper, we chose 3×3 and 5×5 windows pixels for *LZZBP* and different block sizes $w_b = 6, 8, 10, 12$ for histogram matching.

3 Experimental Results

In this section, *ZZPLELD* is evaluated on one illumination variation scenario and two different HFR scenarios, i.e., face sketch vs. photo recognition and NIR image vs. VIS image recognition, respectively, on the existing benchmark databases. For illumination variation scenario, we tested our proposed method on Extended Yale B Face Database [38, 39]. For face sketch vs. photo recognition, we tested the proposed method on the CUHK Face Sketch FERET Database (CUFSF) [21]. CASIA-HFB Face Database [40] is used for NIR face image vs VIS face recognition testing.

At first, proposed *ZZPLELD* is tested on illumination-invariant face recognition. We compared *ZZPLELD* with several other methods, namely LBP, gradient-face [36], TVQI [37], HF+HQ [41], MSLDE [32] on Extended Yale B Face Database. All the methods mentioned above are well tuned according to their respective published papers.

We compared *ZZPLELD* with several state-of-the-art methods, namely PLS [15], CITE [21], MCCA [25], TFSPS [10], KP-RS [23], MvDA [18], G-HFR [30], and LGFP [28] on viewed sketch database. We also compared *ZZPLELD* with several state-of-the-art methods, namely KP-RS, LCKS-CSR [17], and THFM [24] on NIR face image vs VIS face image database. Experimental setups (training and testing samples) and accuracies of the methods mentioned above, except LGFP, are taken from the published papers.

3.1 Rank-1 Recognition Results on Extended Yale B Database

This database contains total 2432 images of 38 subjects under 64 different illumination conditions and in a cropped form with a size of 192×168 pixels. Again, the database is divided into five different subsets according to the illumination angle: Subset 1 ($0°–12°$, 7 images per subject), Subset 2 ($13°–25°$, 12 images per subject), Subset 3 ($26°–50°$, 12 images per subject), Subset 4 ($51°–77°$, 14 images per subject), and Subset 5 ($78°$ and above, 19 images per subject). Figure 4 shows one sample face images under different illumination conditions from the Extended Yale B database and their corresponding *ZZPLELD* images. For the experiment, the image with the most neutral light condition without illumination for each subject from Subset 1 was defined as the gallery, and the remaining images from Subset 1 to Subset 5 were used as query images. A comparison on of the rank 1 accuracy achieved on this database of 38 subjects on the individual subset and after averaged over all subsets is shown in Table 1.

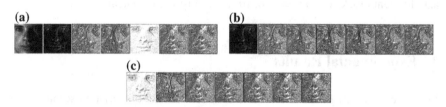

Fig. 4 **a** Sample *ZZPLELD* images after applying 3×3 *LZZBP* on both extremum *LELD* from Extended Yale B Face database. **b** Sample *ZZPLELD* images after applying 5×5 *LZZBP* on maximum *LELD* from Extended Yale B Face database. **c** Sample *ZZPLELD* images after applying 5×5 *LZZBP* on minimum *LELD* from Extended Yale B Face database

Table 1 Rank-1 recognition rates for different methods on different subsets of Extended Yale B database of 38 subjects

State-of-the-art Methods	S1	S2	S3	S4	S5	Avg
LBP	84.81	89.91	80.04	72.18	74.99	80.39
Gradient-face	94.74	**100**	83.33	75.94	84.65	87.73
TVQI	92.28	95.20	90.27	81.39	84.32	88.69
HF+HQ	94.81	98.76	93.18	82.90	84.43	90.82
MSLDE	96.61	**100**	93.20	86.66	89.33	93.16
ZZPLELD	**97.01**	**100**	**94.56**	**90.62**	**90.37**	**94.51**

3.2   Rank-1 Recognition Results on Viewed Sketch Databases

A CUHK Face Sketch FERET (CUFSF) database has been used for the experimental study, which includes 1194 different subjects from the FERET database. For each person, there is a sketch with shape exaggeration drew by an artist when viewing this photo and a face photo with lighting variations. All the frontal faces in the database are cropped manually by setting approximately the same eye levels and resized to 120 × 120 pixels. The proposed method is tested with the existing state-of-the-art methods (PLS, CITE, MCCA, TFSPS, KP-RS, MvDA, G-HFR, and LGFP). Experimental setups and results of other state-of-the-art methods are collected from the published papers. The rank-1 recognition result of proposed *ZZPLELD* on CUFSF database is 96.35% at rank-1, and it is shown in Table 2. Proposed method outperforms other state-of-the-art methods. Figure 5 shows one sample face photo and sketch image from the CUFSF database and their corresponding *ZZPLELD* images.

Table 2 Rank-1 recognition rates for different methods on CUFSF database

State-of-the-art methods	Number of training samples (Subjects)	Number of testing samples (Subjects)	Rank-1 accuracy (%)
PLS	300	894	51.00
CITE	500	694	89.54
MCCA	300	894	92.17
TFSPS	300	894	72.62
KP-RS	500	694	83.95
MvDA	500	694	55.50
G-HFR	500	694	96.04
LGFP	0	1194	94.47
ZZPLELD	0	1194	**96.35**

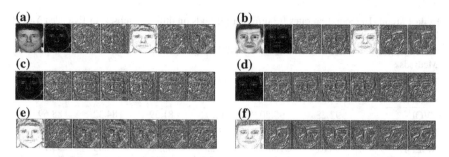

Fig. 5 Sample Sketch and photo images and their corresponding *ZZPLELD* images from CUFSF database. **a** Photo-*ZZPLELD* images after applying 3 × 3 *LZZBP* on both extremum *LELD*, **b** Sketch-*ZZPLELD* images after applying 3 × 3 *LZZBP* on both extremum *LELD*, **c** Photo-*ZZPLELD* images after applying 5 × 5 *LZZBP* on maximum *LELD*, **d** Sketch-*ZZPLELD* images after applying 5 × 5 *LZZBP* on maximum *LELD*, **e** Photo-*ZZPLELD* images after applying 5 × 5 *LZZBP* on minimum *LELD*, **f** Sketch-*ZZPLELD* images after applying 5 × 5 *LZZBP* on minimum *LELD*

3.3 Rank-1 Recognition Results on NIR-VIS CASIA-HFB Database

This database has 200 subjects with probe images captured in the near-infrared and gallery images captured in the visible light. Each and every subject has 4 NIR images and 4 VIS images with pose and expression variations. All the frontal faces in the database are cropped manually by setting approximately the same eye levels and resized to 120 × 120 pixels. This database follows standard evaluation protocols.

The output result is also tested against other state-of-the-art methods (KP-RS, LCKS-CSR, THFM). Table 3 shows the rank-1 accuracy of the proposed method and other state-of-the-art methods. The rank-1 recognition of all those methods, mentioned above, is found from different published papers. The rank-1 recognition accuracy of the proposed method is 99.39%, and it is better than other methods. One sample NIR-VIS pair image from CASIA-HFB database and its different levels of *ZZPLELD* is shown in Fig. 6.

Table 3 Rank-1 recognition rates for different methods on CASIA-HFB database of 200 subjects

State-of-the-art methods	Number of training samples (Subjects)	Number of testing samples (Subjects)	Rank-1 accuracy (%)
KP-RS	133	67	87.80
LCKS-CSR	150	150	81.43
THFM	100	100	99.28
ZZPLELD	0	200	**99.39**

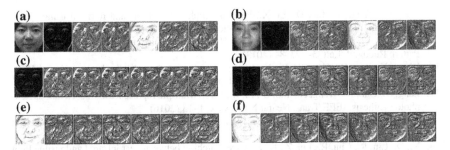

Fig. 6 Sample VIS and NIR images and their corresponding *ZZPLELD* images from CASIA-HFB database, **a** VIS-*ZZPLELD* images after applying 3 × 3 *LZZBP* on both extremum *LELD*, **b** NIR-*ZZPLELD* images after applying 3 × 3 *LZZBP* on both extremum *LELD*, **c** VIS-*ZZPLELD* images after applying 5 × 5 *LZZBP* on maximum *LELD*, **d** NIR-*ZZPLELD* images after applying 5 × 5 *LZZBP* on maximum *LELD*, **e** VIS-*ZZPLELD* images after applying 5 × 5 *LZZBP* on minimum *LELD*, **f** NIR-*ZZPLELD* images after applying 5 × 5 *LZZBP* on minimum *LELD*

4 Conclusion

We have presented a novel modality-invariant feature representation *ZZPLELD* for HFR. It is a combination of *LELD* and *LLZZBP* to boost the performance of HFR. The proposed *LELD* is an illumination-invariant image representation. To capture the local patterns of *LELD*, a novel zigzag binary pattern (*LLZZBP*) is proposed.

Experimental results on illumination variations, sketch-photo and NIR-VIS databases, the proposed method shows the supremacy in rank-1 recognition than other compared methods. The result shows that *ZZPLELD* has a good verification and discriminating ability in heterogeneous face recognition.

ZZPLELD can have promising prospect in other fields of image processing and which worth further investigations.

References

1. Li, S., Chu, R., Liao, S., Zhang, L.: Illumination invariant face recognition using NIR images. IEEE Trans. Pattern Anal. Mach. Intell. **29**(4), 627–639 (2007)
2. Li, S.: Encyclopaedia of Biometrics. Springer (2009)
3. Tang, X., Wang, X.: Face sketch recognition. IEEE Trans. Circuits Syst. Video Technol. **14**(1), 50–57 (2004)
4. Chen, J., Yi, D., Yang, J., Zhao, G., Li, S., Pietikainen, M.: Learning mappings for face synthesis from near infrared to visual light images. In: Proceedings of IEEE International Conference on Computer Vision and Pattern Recognition, pp. 156–163 (2009)
5. Gao, X., Zhong, J., Li, J., Tian, C.: Face sketch synthesis algorithm on e-hmm and selective ensemble. IEEE Trans. Circuits Syst. Video Technol. **18**(4), 487–496 (2008)
6. Wang, X., Tang, X.: Face photo-sketch synthesis and recognition. IEEE Trans. Pattern Anal. Mach. Intell. **31**(1), 1955–1967 (2009)
7. Li, J., Hao, P., Zhang, C., Dou, M.: Hallucinating faces from thermal infrared images. In: Proceedings IEEE International Conference on Image Processing, pp. 465–468 (2008)

8. Gao, X., Wang, N., Tao, D., Li, X.: Face sketchphoto synthesis and retrieval using sparse representation. IEEE Trans. Circuits Syst. Video Technol. **22**(8), 1213–1226 (2012)
9. Wang, N., Li, J., Tao, D., Li, X., Gao, X.: Heterogeneous image transformation. Elsevier J. Pattern Recognit. Lett. **34**, 77–84 (2013)
10. Wang, N., Tao, D., Gao, X., Li, X., Li, J.: Transductive face sketch-photo synthesis. IEEE Trans. Neural Netw. **24**(9), 1364–1376 (2013)
11. Peng, C., Gao, X., Wang, N., Tao, D., Li, X., Li, J.: Multiple representation-based face sketch-photo synthesis. IEEE Trans. Neural Netw. xx x, 1–13 (2016)
12. Lin, D., Tang, X.: Inter-modality face recognition. In: Proceedings of European Conference on Computer Vision, pp. 13–26 (2006)
13. Yi, D., Liu, R., Chu, R., Lei, Z., Li, S.: Face matching between near infrared and visible light images. In: Proceedings of International Conference on Biometrics, pp. 523–530 (2007)
14. Lei, Z., Li, S.: Coupled spectral regression for matching heterogeneous faces. In: Proceedings of IEEE International Conference on Computer Vision and Pattern Recognition, pp. 1123–1128 (2009)
15. Sharma, A., Jacobs, D.: Bypassing synthesis: Pls for face recognition with pose, low-resolution and sketch. In: Proceedings of IEEE International Conference on Computer Vision and Pattern Recognition, pp. 593–600 (2011)
16. Mignon, A., Jurie, F.: Cmml: a new metric learning approach for cross modal matching. In: Proceedings of Asian Conference on Computer Vision, pp. 1–14 (2012)
17. Lei, Z., Liao, S., Jain, A.K., Li, S.Z.: Coupled discriminant analysis for heterogeneous face recognition. IEEE Trans. Inf. Forensics Secur. **7**(6), 1707–1716 (2012)
18. Kan, M., Shan, S., Zhang, H., Lao, S., Chen, X.: Multi-view discriminant analysis. IEEE Trans. Pattern Anal. Mach. Intell. **38**(1), 188–194 (2016)
19. Liao, S., Yi, D., Lei, Z., Qin, R., Li, S.: Heterogeneous face recognition from local structure of normalized appearance shared representation learning for heterogeneous face recognition. In: Proceedings of IAPR International Conference on Biometrics (2009)
20. Klare, B.F., Li, Z., Jain, A.K.: Matching forensic sketches to mug shot photos. IEEE Trans. Pattern Anal. Mach. Intell. **33**(3), 639–646 (2011)
21. Zhang, W., Wang, X., Tang, X.: Coupled information-theoretic encoding for face photo-sketch recognition. In: Proceedings of IEEE International Conference on Computer Vision and Pattern Recognition, pp. 513–520 (2011)
22. Bhatt, H.S., Bharadwaj, S., Singh, R., Vatsa, M.: Memetically optimized mcwld for matching sketches with digital face images. IEEE Trans. Inf. Forensics Secur. **7**(5), 1522–1535 (2012)
23. Klare, B.F., Jain, A.K.: Heterogeneous face recognition using kernel prototype similarities. IEEE Trans. Pattern Anal. Mach. Intell. **35**(6), 1410–1422 (2013)
24. Zhu, J., Zheng, W., Lai, J., Li, S.: Matching nir face to vis face using transduction. IEEE Trans. Inf. Forensics Secur. **9**(3), 501–514 (2014)
25. Gong, D., Li, Z., Liu, J., Qiao, Y.: Multi-feature canonical correlation analysis for face photo-sketch image retrieval. In: Proceedings of ACM International Conference on Multimedia, pp. 617–620 (2013)
26. Roy, H., Bhattacharjee, D.: Heterogeneous face matching using geometric edge-texture feature (getf) and multiple fuzzy-classifier system. Elsevier J. Appl. Soft Comput. **46**, 967–979 (2016)
27. Roy, H., Bhattacharjee, D.: Local-gravity-face (LG-face) for illumination-invariant and heterogeneous face recognition. IEEE Trans. Inf. Forensics Secur. **11**(7), 1412–1424 (2016)
28. Roy, H., Bhattacharjee, D.: Face sketch-photo matching using the local gradient fuzzy pattern. IEEE J. Intell. Syst. **31**(3), 30–39 (2016)
29. Roy, H., Bhattacharjee, D.: Face sketch-photo recognition using local gradient checksum: LGCS. Springer Int. J. Mach. Learn. Cybern. **xx**(x), 1–13 (2016)
30. Peng, C., Gao, X., Wang, N., Li, J.: Graphical representation for heterogeneous face recognition. IEEE Trans. Pattern Anal. Mach. Intell. **xx**(x), 1–13 (2016)
31. Sinha, P., Balas, B., Ostrovsky, Y., Russell, R.: Face recognition by humans: Ninetenn results all computer vision researchers should know about. Proc. IEEE, **94** (2006)

32. Lai, Z., Dai, D., Ren, C., Huang, K.: Multiscale logarithm difference edgemaps for face recognition against varying lighting conditions. IEEE Trans. Image Process. **24**(6), 1735–1747 (2015)
33. Ojala, T., Pietikinen, M., Menp, T.: Multiresolution gray-scale and rotation invariant texture classification with local binary patterns. IEEE Trans. Pattern Anal. Mach. Intell. **24**(7), 971–987 (2002)
34. Land, E.H., McCann, J.J.: Lightness and Retinex theory. J. Opt. Soc. Am. **61**(1), 1–11 (1971)
35. Horn, B.K.P.: Robot Vision. MIT Press, Cambridge (2011)
36. Zhang, T., Tang, Y.Y., Fang, B., Shang, Z., Liu, X.: Face recognition under varying illumination using gradientfaces. IEEE Trans. Image Process. **18**(11), 2599–2606 (2009)
37. An, G., Wu, J., Ruan, Q.: An illumination normalization model for face recognition under varied lighting conditions. Elsevier J. Pattern Recognit. Lett. **31**, 1056–1067 (2010)
38. Belhumeur, P., Georghiades, A., Kriegman, D.: From few to many: illumination cone models for face recognition under variable lighting and pose. IEEE Trans. Pattern Anal. Mach. Learn. **23**(6), 643–660 (2001)
39. Lee, K.C., Ho, J., Kriegman, D.: Acquiring linear subspaces for face recognition under variable lighting. IEEE Trans. Pattern Anal. Mach. Learn. **27**(5), 684–698 (2005)
40. Li, S.Z., Lei, Z., Ao, M.: The hfb face database for heterogeneous face biometrics research. In: Proceedings of IEEE International Workshop on Object Tracking and Classification Beyond and in the Visible Spectrum, Miami (2009)
41. Fan, C.N., Zhang, F.Y.: Homomorphic filtering based illumination normalization method for face recognition. Elsevier J. Pattern Recognit. Lett. **32**, 1468–1479 (2011)

Part III
Pattern Recognition

Automatic Extraction and Identification of *Bol* from *Tabla* Signal

Rajib Sarkar, Ankita Singh, Anjishnu Mondal and Sanjoy Kumar Saha

Abstract In Indian classical music, *tabla* is the most widely used rhythmic instrument. The instrument has two drums. By striking either of the drums, a *bol* is produced and it forms the basic component of *tala* (rhythm). In this work, *bol*s are automatically extracted from *tabla* signal. Subsequently, features are extracted and used for *bol* identification. Ideally, a *bol* follows attack-decay-sustain-release (ADSR) model. A *bol* has a characteristic rise in the initial attack stage, after which it decays to reach a steady energy level. It sustains that level and, finally, releases the energy. Proposed segmentation methodology exploits this phenomenon to extract the *bol*s. Once the *bol* segments are extracted, low-level spectral features are computed and used for classification. Multilayer perceptron network is used for *bol* identification. Experiment is successfully carried out with the signals of recitals by different players and also at different tempo. The result shows that proposed methodology performs quite well on diverse collection. Segmentation and identification of *bol*s can act as the foundation for the applications like transcript generation, *tala* identification.

Keywords Tabla signal · Bol segmentation · ADSR model · Bol identification

1 Introduction

Tabla is considered as the queen of all percussion instruments. It is the most important instrument which is used to maintain the rhythm in Indian classical music. It consists

R. Sarkar (✉) · A. Singh · A. Mondal · S. K. Saha
CSE Department, Jadavpur University, Kolkata 700032, West Bengal, India
e-mail: rjbskar@gmail.com

A. Singh
e-mail: singh.ankita91@outlook.com

A. Mondal
e-mail: anjishnu@outlook.com

S. K. Saha
e-mail: sks_ju@yahoo.co.in

© Springer Nature Singapore Pte Ltd. 2018
R. Chaki et al. (eds.), *Advanced Computing and Systems for Security*,
Advances in Intelligent Systems and Computing 666,
https://doi.org/10.1007/978-981-10-8180-4_9

139

of two drums. The bass drum called *bayan* and the treble drum is called *dayan*. The drums have different regions responsible for producing different tones. The playing technique involves the complex use of palm and fingers in various configurations to create a wide variety of different sounds and rhythms. The heel of the hand is used to apply pressure or in a sliding motion on the larger drum so that the pitch is changed during decay of the sound. Thus, the combination of the use of fingers and palm along with the region of the drum(s) gives rise to different strokes/*bols*.

Bols are broadly categorized as either single stroke or as combined stroke type. Combined strokes are produced when two unique single strokes on both the drums are played simultaneously. For understanding, few examples of these two types of *bols* are discussed as follows.

Khe is a single stroke played on the *bayan*. It is also known as *Ka*. It is played by lightly slapping a particular region on *bayan* with a completely flat palm. All the five fingers make contact with the drum. *Na* is a single stroke played on the *dayan*. It is also known as *Ta*. It is played by tapping a particular region with the index finger and with a relatively high pressure.

Combination of single strokes is often used to generate combined *bols*. Being a combination of both the drums, the sound produced is a mixture of treble as well as bass. *Dha* is a combined stroke obtained when *Na* on *dayan* and *Ghe* on *bayan* are played together. *Dhin* is generated by playing *Teen* on *dayan* and *Ghe* on *bayan*.

Bols are the most fundamental unit in *tabla* signal. Sequence of *bols* forms the basic cycle of various *talas*. Hence, an automated system for segmentation of the *tabla* signal into *bols* and their identification act as the fundamental step for the applications like automated transcript generation, *tala* detection etc.

The paper is organized as follows. Section 2 presents the survey on past work. Proposed methodology is elaborated in Sect. 3. Experimental result and concluding remarks are put in Sects. 4 and 5, respectively.

2 Past Work

Identification of *bol* from *tabla* signal is preceded by the task of segmenting the signal into units representing the individual *bols*. Hence, this section includes the review on segmentation. Most of the early works deal with Western music where drum is used as the rhythm instrument. Efforts are mostly directed toward the applications like tempo detection, beat tracking. In this context, detecting the onset is important as it denotes the start of a beat in the drum signal. As onset detection is also utilized in extracting the *bols* from *tabla* signal, it is worth to go through such works.

Bello et al. [1] proposed an onset detection method that combines energy difference-based function and phase-based approaches. The energy-based approach brings out strong percussive onsets while the phase approach brings out the soft onsets. Further improvement for onset detection was achieved in [2] when preprocessing is applied to decompose the signal into multiple bands and different reduction methods are applied on the bands to determine the onsets. Dixon [3] proposed further improvements on the work done in [1, 2] by introducing the use of weighted phase deviation, normalized weighted phase deviation, and half-wave rectified complex domain onset detection functions. Similar approach was also adopted by Grosche and Mller [4].

Scheirer in his work [5] pointed out that an audio signal needed to be divided into multiple bands and sub-band envelopes, needed to be processed separately for accurate rhythm detection. The concept of psycho-acoustic knowledge described by Scheirer [5] was implemented by Klapuri [6]. Foote [7] presented the concept of novelty curve and based on the same proposed a methodology for onset detection [8]. It relies on similarity between the audio frames. This approach is well in use in various applications like beat tracking, tempo detection.

Gillet and Richard [9] worked on *bol* identification. Segmentation is achieved by applying a low threshold on low frequency envelope and onset detection. Spectral-based features are used as the descriptors. Identification is done based on hidden Markov model (HMM). Chordia [10] presented a methodology for *bol* extraction and detection. Similar to the work of Grosche and Mller [4], the amplitude and phase information are combined to identify the onsets. Strokes are then extracted based on the steady-state assumptions. As in the work of Gillet and Richard [9], spectral features such as centroid, skewness, kurtosis, rolloff are used as descriptors for *bol* identification. Descriptors also include mel-frequency cepstrum coefficient (MFCC), temporal centroid, attack-time, and zero crossing rate. Four different classifiers like multivariate Gaussian model, feed-forward neural network, probabilistic neural network, and tree classifier are used. Experiment is done on three datasets which correspond to three players. The recognition rates are found to be highest whenever the training and testing were performed on the same dataset. The results are poor when training and testing were performed on mixture of the three datasets. Thus, it is not so capable of handling the variety. The similar system is further utilized to develop a real-time recognition system [11].

Marius-Miron developed a system for automatic recognition of *tala* [12] that extended the work in [11]. Bayesian information criteria are used at first-level segmentation. Subsequently, the outcome is refined based on density of inter-onset interval, correlation of inter-onset histogram. Each *bol* is modeled using Gaussian mixture model (GMM), and the models are used in classifying the segmented units. Onset detection influences the identification outcome. The reported result shows both miss and over-detection of onsets. The dataset used in experiment is very limited and tests are performed on only four *bol*s.

It is observed that very few works have been reported which focused on *tabla* signal. Moreover, mere use of the onset detection techniques is not sufficient for extracting the *bol*s from *tabla* signal. Unlike the beats in a drum signal, *bol*s are of wide variety. Non-uniformity of the individual *bol* makes the task further challenging. Thus, segmentation and subsequent identification of *bol*s in *tabla* signal demand attention. A robust recognition system will act as the fundamental step toward the application like automated transcript generation, *tala* (rhythm) identification, self-learning, and evaluation system for playing *tabla*.

3 Proposed Methodology

Tabla is the widely used rhythm instrument used in Indian classical music, and *bol* is the basic alphabet in the grammar of *Tabla*. In this work, our goal is to extract and then identify the *bol*s in *tabla* signal. Unlike the drums, recital of *tabla* reflects a cyclic pattern that repeats. The cycle consists of a sequence of *bol*s and gives rise to different *tala*. An optimal strategy for extracting and identifying the *bol*s will be of immense use for higher level applications like to transcript generation, *tala* detection, and evaluation of a learner. Proposed methodology consists of two major modules, namely *bol segmentation* and *bol identification*. The segmentation is achieved based on attack-decay-sustain-release (ADSR) model, and thereafter low-level descriptors are used in identifying the extracted *bol*s.

3.1 Bol Segmentation

Proposed segmentation methodology consists of two steps. At first, it utilizes the concept of onset detection [2] for approximating the *bol* segment, and thereafter, it is refined to obtain the final segmentation on the input signal. Major contribution in segmentation mostly lies in the refinement. Two major modules are as follows.

- Approximating the *bol* location
- Extracting the *bol* segment.

The first stage detects the onset location on the processed version of the original signal which is to be mapped on the original signal. The direct mapping provides an approximation. The second stage provides the improvement for localizing the *bol* in the original signal.

3.1.1 Approximating *Bol* Location

A beat in case of drum or a *bol* played on *tabla* is expected to follow attack-decay-sustain-release (ADSR) model as shown in Fig. 1. The attack phase extends from the start of the *bol* to the instance of reaching maximum energy. After that, energy decays for certain duration till it stabilizes. Sustained state has considerably steady energy level. Then energy is gradually dropped denoting the release period. Attack and the decay phases together form the transient stage reflecting the change in energy level. Onset denotes the start of a beat or *bol*. Thus, according to ADSR model onset corresponds to the instantiation of attack phase. To detect the onsets, we rely on well-tried mechanism based on novelty curve.

Novelty curve reflects the change of property in the input signal. By capturing and analyzing the variation, it tries to find out the onset. Audio signal is oscillatory in nature. Hence, change detection by differentiating the time domain signal is not

Fig. 1 Attack-decay-sustain-release (ADSR) model

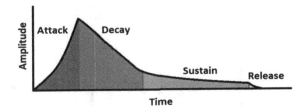

fruitful. It needs to be transformed in an intermediate form that reflects the transient characteristics of the signal. Such a transformation function is referred as detection/novelty function. The transformed signal is highly sub-sampled in comparison to the original signal, and in a broad sense it is the novelty curve. The peaks in the novelty curve denote the point of high change in property and considered as the onset of a note. This is because initiation of attack phase is likely to result into maximum change in property. Thus, the major issues are *selecting the novelty function, i.e., transformation* and *choosing the peaks*. These are detailed as follows.

Transforming the signal: Signal undergoes variation in energy during the transient phase. Hence, in our work, energy is considered as the property for novelty curve generation. Spectral features are considered for transformation. The signal is first sub-sampled and then divided into number of fixed size frames. In our experiment, frame size is taken as 256 samples. For each frame, short-time Fourier transform (STFT) is computed. Based on the sequence of STFTs, each band is analyzed separately. It is likely that the transient characteristics are more pronounced in the higher frequency bands. For each band, spectral difference over consecutive frames is computed and it plays the role of detection function. It reflects the variation of spectral energy over time in the individual band. Corresponding elements of all the bands are averaged to obtain the novelty curve depicting the change in spectral energy over time.

Selecting the peaks: Normally, local maxima in the novelty curve are taken as the peaks. But it may give rise to spurious peaks. The elements of novelty curve are normalized by subtracting mean and dividing it by the maximum deviation. Figure 2 shows a sample *tabla* signal and corresponding normalized novelty curve. On this normalized curve, local maxima above a certain threshold are considered as the peaks. In our work, threshold is taken as $\mu - 0.5 \times \sigma$, where μ and σ stand for average and standard deviation of local maxima values. Finally, based on sub-sampling factor and frame size, time instance of the detected peaks are mapped to the original signal. This denotes the approximate position of the onset/start position of a *bol* in the original signal.

3.1.2 Extracting *Bol*

Unlike the drum, in *tabla* recital *bol*s are played continuously one after another. Moreover, *bol*s vary widely. Palm and finger activities are different for different *bol*.

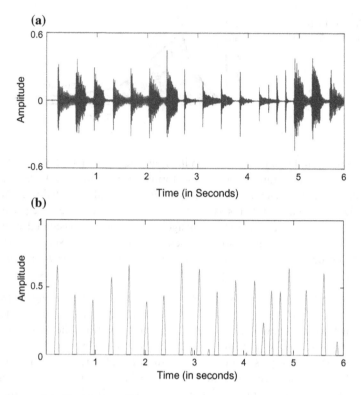

Fig. 2 **a** A sample *tabla* signal and **b** corresponding novelty curve

Moreover, they are played using different regions in the instrument. All these results in variation in terms of energy and frequency characteristics of the *bol*s. Depending on the *bol* sequence played for a *tala* and also on the tempo being used, phases of ADSR may not be well reflected for individual *bol*s. Thus, ADSR pattern for a *bol* in isolation and same for the *bol* in continuous stream are further different as it bears the impression of its neighbors. Audio signal is additive and very often release phase of preceding *bol* and attack phase of following one overlap. All these specialties account for the shortcomings of conventional onset detection mechanism in the context of locating the start of a *bol*. But it provides an estimate and one can look into the neighborhood of such points for precise detection. However, for the application like tempo detection, such approximated localization serves the purpose. But, for *bol* identification, the impreciseness may be costly. The onset detected at the first stage certainly corresponds to an instance in transient phase. Based on that reference point, location is further refined. The refinement process consists of two steps as follows.

- Determine the attack-decay junction point
- Backtrack to refine the position of onset

Determine the attack-decay junction point: Because of the variety of *bols* and specialty of *tabla* recital, novelty curve-based onset detection mechanism as presented in Sect. 3.1.1 identifies a point (let us refer it as candidate point) in the transient phase. To be more specific, it falls in attack phase for most of the cases. Depending on the additive effect of consecutive *bols* of different nature, the candidate point may actually be in decay phase also. This further adds to the problem.

In ADSR model, at the junction of attack and decay phase energy is maximum. We try to find out this junction point. It lies within a neighborhood of candidate point. As the candidate point may lie in attack or decay phase, the neighborhood for searching the maximum energy point must include both forward and backward region with respect to the candidate point. Search range is determined based on the interval between the candidate points. Suppose we want to refine the ith candidate point. Let t be an estimate for the corresponding *bol* duration and $t = t_{ci+1} - t_{ci}$, where t_{cj} stands for the time stamp for j-th candidate point. All the four phases are not of equal duration, and it is more likely that all the phases are not clearly reflected in the continuous recital. Based on this understanding, we have experimentally determined $t/3$ as the neighborhood in timescale for finding maximum energy point. Search space is centered at the candidate point in the rectified signal and spans over determined time period. Within the defined space, instance with maximum magnitude corresponds to the attack-decay junction point.

Backtrack to refine the position of onset: Starting from the attack-decay junction point, we backtrack to determine the attack phase as far as possible. At this stage, also we work with rectified signal. Let t_{jpi} and t_{ci} are the time stamps for junction point and candidate point for ith *bol*. In our backtracking process, we need to ensure that t_{ci} is in the attack phase. In case it is in decay phase, i.e., $t_{ci} > t_{jpi}$, we modify t_{ci} as follows.

$$diff = t_{ci} - t_{jpi}$$

$$t_{ci} = t_{jpi} - diff$$

Starting from t_{jpi}, we backtrack till t_{ci}, to determine the time points with decreasing energy. There may be minor fluctuations and hence instead of considering energy at individual points, average energy over the small non-overlapping windows can be considered. Even after windowing, some of the intermediate values may violate the decreasing trends and those are ignored. Thus, from the said interval set of monotonically decreasing energy points are identified. A straight line is fitted trough the points by least square regression, and it is extrapolated till the energy drops to zero and t_{ai} be the corresponding time stamp for zero energy. Within the interval $[t_{ai}, t_{jpi}]$, the time stamp corresponding to minimum energy is taken as the final onset or start of the ith *bol*. Refinement process is applied for all the candidate points. Let $t_1, t_2, ..., t_n$ are the corrected onset points. Then, the input signal within the interval $[t_i, t_{i+1}]$ represents the ith *bol*. Figure 3 shows one sample output of *bol* segmentation.

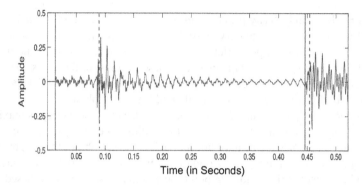

Fig. 3 Segmentation output–dashed line denotes the *bol* segment determined using novelty curve and solid line denotes the same determined by proposed methodology

It is observed that the boundary detected by the proposed methodology is closer than that by novelty curve only.

3.2 Bol Identification

Playing style of the *bols* vary. Some are played on *bayan* and some are on *dayan*. Regions of *tabla* and number of fingers used in playing different *bols* are also not same. All these contribute toward the variation in energy and frequency for the *bols*. These observations have motivated us to consider spectral features for designing the descriptors for the extracted *bols*. Fast Fourier Transform (FFT) is applied on the extracted signal to obtain the frequency spectrum and thereafter following features are computed.

- **Spectral Centroid**: It is a gross summarization of the frequency spectrum that indicates where the center of mass of the spectrum lies. It has a robust connection with the impression of brightness of an audio. It is computed as the weighted mean of the frequencies present in the spectrum, with their magnitudes as the weights.

$$centroid = \frac{\sum_{n=0}^{N-1} f(n)x(n)}{\sum_{n=0}^{N-1} x(n)}$$

  where $x(n)$ represents the weighted magnitude of frequency bin number n, and $f(n)$ represents the center frequency of the bin.

- **Spectral Flatness**: A high spectral flatness indicates that the spectrum has a similar amount of power in all spectral bands, while a low spectral flatness indicates that the spectral power is concentrated in relatively small number of bands. It is computed as ratio of geometric mean and arithmetic mean of the power spectrum.

$$flatness = \sqrt[N]{\frac{\prod_{n=0}^{N-1} x(n)}{\frac{\sum_{n=0}^{N-1} x(n)}{N}}}$$

where $x(n)$ represents the magnitude of bin number n.

- **Spectral Skewness**: It provides the skewness in spectrum distribution. The duration (time scale) and intensity (energy) distribution during attack-decay-sustain-release stages of a *bol* are likely to influence the measure.
- **Spectral Kurtosis**: Spectral kurtosis provides a means of determining which frequency bands contains a signal of maximum impulsivity. It is computed as

$$x = \frac{\mu_4}{\sigma_4} = \frac{\Sigma[(X - \mu)^4]}{\sigma[(X - \mu)^2]^2}$$

where μ_4 is the fourth moment about the mean and σ is the standard deviation. The value of spectral kurtosis is close to zero for Gaussian noise and large for signals containing series of short transients.

- **Spectral Rolloff**: Spectral rolloff point is defined as the Nth percentile of the power spectral distribution. In this work, N is taken as 85%. The rolloff point is the frequency below which the $N\%$ of the magnitude distribution is concentrated.
- **Mel-Frequency Cepstral Coefficients (MFCC)**: MFCCs are the results of a cosine transform of the real logarithm of the spectrum expressed on a mel-frequency scale. It resembles human perception of hearing.

We have used multilayer perceptron (MLP) classification algorithm incorporated in Weka [13] to identify segmented *bols*. It consists of sets of adaptive weights, i.e., connection strength between nodes, which are tuned by a learning algorithm. MLP has three layers. The first layer is input layer and the last layer is called output layer. Apart from the input and output layers, there exists one or more intermediate hidden layers. In our experiment, only one hidden layer is considered. Number of nodes in input layer is equal to the dimension of input feature vector. Number of nodes in output layer is equal to the target classes.

4 Experimental Results

Bols are the building block of all *talas* in *tabla* recital. We have prepared a dataset by recording the recitals of number of artists with different skill-set. There are number of clips amounting to 75 min duration. Four different *talas*, namely *Teental*, *Dadra*, *Kaharba*, and *Ektal* are considered. Again the playing style varies depending on *gharana* (school of practice). For each *tala*, clips corresponding to three different *gharanas* are present in the collection. Altogether seven artistes have played. *Tabla* was also tuned at different frequencies. Tempo varied from 80 to 180 BPM. Sampling rate is 44.1 kHz. Different *tablas* are used for recordings and those are tuned at different frequencies. Thus, the collection reflects ample variations in different aspects.

Table 1 Grammar of talas

Tala	Grammar of tala
Teental	Dha, Dhin, Dhin, Dha, Dha, Dhin, Dhin, Dha, Na, Teen, Teen, Na, Teh Te, Dhin, Dhin, Dha
Kaharba	Dha, Ghe, Teh, Te, Na, Khe, Dhin, Na
Dadra	Dha, Dhin, Na, Na, Teen, Na
Ektal	Dhin, Dhin, DhaGe, TiTaKiTa, Tu, Na, Ka, Ta, DhaGe, TiTaKiTa, Dhi, Na

Table 2 Identification of attack-decay junction points

Tala	# of junctions	# of junctions	Detection
	Expected	Correctly identified	Accuracy (in %)
Teentala	1268	1268	100.00
Ektala	1848	1848	100.00
Dadra	911	911	100.00
Kaharba	786	786	100.00

Tala consists of a sequence of *bols* which is repeated. This is the specialty of *tabla* making it different from drum. The core cycle of *bols* for four *talas* is shown in Table 1. As discussed earlier, *bols* reflect variation. The *bols Na* and *Khe* appearing consecutively in *Kaharba tala* made segmentation process further difficult. *Na* is of high duration and emphasized. On the contrary, *Khe* is of very low duration and relatively feeble in terms of energy. *Ghe* and *Teh* are also of low energy. Thus in a *tala*, co-occurrence of varying *bols* makes the segmentation critical.

Based on novelty curve, candidate points are detected which lies in transient phase. Utilizing this, we first determine the attack-decay junction points and, finally, the corrected onsets are identified. Identified junctions and onsets are mapped onto the rectified signal to judge the correctness manually. The performance of the junction point detection methodology is summarized in Table 2. Table 3 shows the accuracy of onset detection achieved after refinement. Results indicate that the performance of proposed methodology is satisfactory. Figure 4 shows a sample result of *bol* segmentation. The vertical line denotes the manual demarcation between the two *bols*. It is observed that finding the correct onset is more challenging in comparison the attack-decay junction point detection. The problem mostly arises due to the co-occurrence of weak and strong *bols*, low and high duration *bols*, or both.

In order to carry out the performance analysis of the *bol* identification methodology, we have first prepared the ground truth of extracted *bols*. A number of experts have listened the *tabla* recital recordings as well as the extracted *bol* segments and identified them. The segments for which the listeners agreed form the ground-truth collection. As learning-based methodology is used for identification, we have considered a subset of *bols* having a sizable number of samples. Ten-fold cross-validation is applied to measure the classification accuracy, and it has been shown in Table 4.

Table 3 Identification of *bol* onset (after refinement)

Tala	# of bols	# of correctly Detected onsets	Detection Accuracy (in %)
Teentala	1268	1209	95.35
Ektala	1848	1788	96.75
Dadra	911	895	98.24
Kaharba	786	701	89.19
Overall	4813	4593	95.43

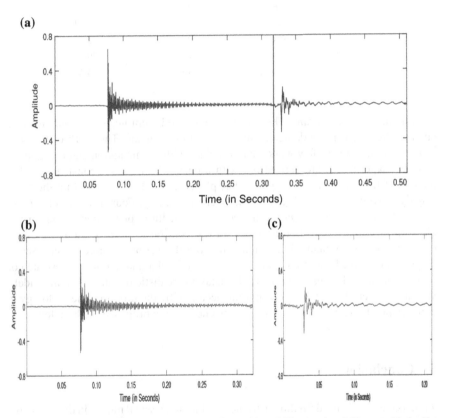

Fig. 4 Segmentation output: **a** Two consecutive *bols*—*Na* and *Khe* together, **b** Extracted *bol*—*Na*, and **c** Extracted *bol*—*Khe*

It is observed that despite of small number of samples the performance is quite satisfactory.

As discussed in Sect. 2, most of the early works evolve around Western music where drum is used as rhythmic instrument. Beats are played in a linear fashion and tracking of the beats is the major activity in rhythm analysis. In Indian music, *tabla* is the mostly used rhythmic instrument. *Bols* of different kind appearing in

Table 4 Identification of Bols

Bol name	# of extracted	Detection
	Segment	Accuracy (%)
Dha	648	85.96
Dhin	1034	86.27
Na	1117	87.02
Tin	310	85.16
Tu	154	74.03
Ka	154	81.82
Ghe	99	72.70
Te	196	80.61
Khe	98	75.51
Ta	154	79.87
Overall	3964	84.56

a cyclic pattern form a *tala*. Thus, segmentation and identification of *bol*s is quite challenging and it is fundamental step for rhythm analysis. The work of Chrodia [10] is one of the very few works that has dealt with segmentation and recognition of *bol*s. Experiments are carried out with a customized dataset consisting of *bol* recitals of three players. Each player has played a fixed set of *bol*s (not the *tala*s usually played). In general, the dataset lacks in variety. Four different classifiers are used for recognition. On the mixed dataset of three plays, depending on the classifier average recognition accuracy varies from 66 to 83%. Although a direct comparison was not made, it may noted that our dataset offers more variety. Seven players have played the actual *tala*s, also in different gharanas (styles). These are not focused on the individual *bol*s. Thus, the dataset is realistic in nature. On this widely varying dataset, the achieved recognition accuracy by the proposed methodology is comparable. It clearly indicates the effectiveness of the proposed methodology.

5   Conclusion

Tabla is the mostly used rhythm instrument in Indian classical music. In this work, we have presented a novel scheme for automatic extraction of the *bol*s from *tabla* recital that derives the motivation from attack-decay-sustain-release (ADSR) model. *Bol*s of wide variety are played continuously. Hence, novelty curve-based scheme as used in beat tracking in Western music does not provide optimal segmentation. We have proposed a refinement over the outcome of conventional novelty curve-based onset detection method. It results into the improvement in determining the onsets. Low-level spectral features are computed from the extracted segments to form the input feature vector for a neural network classifier. Experiment is carried out with a wide variety of collection and the performance of the proposed methodology is satisfactory.

In future, the work can be extended to develop online transcript generation system and detecting *tala* of a *tabla* recital.

References

1. Bello, J.P., Duxbury, C., Davies, M., Sandler, M.: On the use of phase and energy for musical onset detection in the complex domain. IEEE Signal Process. Lett. **11**(6), 553–556 (2004)
2. Bello, J.P., Daudet, L., Abdallah, S., Duxbury, C., Davies, M., Sandler, M.B.: A tutorial on onset detection in music signals. IEEE Trans. Speech Audio Process. **13**(5), 1035–1047 (2005)
3. Dixon, S.: Onset detection revisited. In: Proceedings of the 9th International Conference on Digital Audio Effects, vol. 120, pp. 133–137 (2006)
4. Grosche, P., Mller, M.: Extracting predominant local pulse information from music recordings. IEEE Trans. Audio Speech Lang. Process. **19**(6), 1688–1701 (2011)
5. Scheirer, E.: Tempo and beat analysis of acoustic musical signals. **103**(1), 588–601 (1998)
6. Klapuri, A.: Sound onset detection by applying psychoacoustic knowledge. In: Proceedings of the IEEE International Conference of Acoustics, Speech and Signal Processing, Washington, DC, USA, vol. 6, pp. 115–118 (1999)
7. Foote, J.: Visualizing music and audio using self-similarity. In: Proceedings of the Seventh ACM International Conference on Multimedia (Part 1). MULTIMEDIA '99, pp. 77–80. ACM, New York, NY, USA (1999)
8. Foote, J.: Automatic audio segmentation using a measure of audio novelty. In: IEEE International Conference on Multimedia and Expo (I), pp. 452–455. IEEE Computer Society (2000)
9. Gillet, O., Richard, G.: Automatic labelling of tabla signals (2003)
10. Chordia, P.: Segmentation and recognition of tabla strokes. In: ISMIR, pp. 107–114 (2005)
11. Chordia, P., Rae, A.: Tabla gyan: a system for realtime tabla recognition and resynthesis. In: ICMC (2008)
12. Miron, M.: Automatic detection of hindustani talas. Master's thesis, Universitat Pompeu Fabra, Barcelona, Spain (2011)
13. Witten, I.H., Frank, E., Hall, M.A., Pal, C.J.: Data Mining: Practical Machine Learning Tools And Techniques. Morgan Kaufmann (2016)

Optimum Circle Formation by Autonomous Robots

Subhash Bhagat and Krishnendu Mukhopadhyaya

Abstract This paper considers a constrained version of the *circle formation* problem for a set of asynchronous, autonomous robots on the Euclidean plane. The *circle formation* problem asks a set of autonomous, mobile robots, initially having distinct locations, to place themselves, within finite time, at distinct locations on the circumference of a circle (not defined a priori), without colliding with each other. The *constrained circle formation* problem demands that in addition the maximum distance moved by any robot to solve the problem should be minimized. A basic objective of the optimization constrain is that it implies energy savings of the robots. This paper presents results in two parts. First, it is shown that the constrained circle formation problem is not solvable for oblivious asynchronous robots under *ASYNC* model even if the robots have rigid movements. Then the problem is studied for robots which have $O(1)$ bits of persistent memory. The initial robot configurations, for which the problem is not solvable in this model, are characterized. For other configurations, a distributed algorithm is presented to solve the problem for asynchronous robots. Only one bit of persistent memory is needed in the proposed algorithm.

Keywords Swarm robots · Asynchronous · Circle formation
Robots with persistent lights

1 Introduction

A *robot swarm* consists of small, autonomous, indistinguishable, inexpensive mobile robots. Robots in such a distributed system work cooperatively to accomplish some common task which cannot be done by a single large robot. The robots are autonomous (they work without any centralized control), homogeneous (all of them

S. Bhagat (✉) · K. Mukhopadhyaya
Advanced Computing and Microelectronics Unit,
Indian Statistical Institute, Kolkata, India
e-mail: sbhagat_r@isical.ac.in

K. Mukhopadhyaya
e-mail: krishnendu@isical.ac.in

© Springer Nature Singapore Pte Ltd. 2018
R. Chaki et al. (eds.), *Advanced Computing and Systems for Security*,
Advances in Intelligent Systems and Computing 666,
https://doi.org/10.1007/978-981-10-8180-4_10

153

have same capabilities) and anonymous (they are indistinguishable by their appearances and nature). All of them execute the same algorithm. In general, robots lack explicit communication capability. The robots can implicitly communicate with each other by sensing the positions of other robots in the system, using endowed sensors. The system does not have any global coordinate axes. Each robot owns a local coordinate system having origin at its current position. The local coordinate systems of two different robots may have different directions and orientations of coordinate axes and unit distances. In general, the robots do not remember any piece of information of their previous computational cycles, i.e. they are oblivious.

A robot has one of the two states at any point of time: active or inactive. Initially, all robots are inactive. On activation, a robot executes its computational cycle consisting of *Look–Compute–Move* phases. During the *Look* phase, a robot takes snapshot of its surrounding environment, using its sensing capability, to obtain the positions of other robots. Considering the input from the *Look* phase, the *Compute* phase outputs a destination point for the robot. Finally, the robot moves towards the destination point during the *Move* phase. An idle robot remains silent without performing any course of action. Robots may be endowed with some additional capabilities or may have some common agreements in order to solve different coordination problems. The memory model assumes that the robots possess a constant amount of additional persistent memories to remember their current states. The implementation of such persistent memory is done by externally visible lights. These lights use a constant number of colours. The colours are predefined to indicate different states of the robots [3, 8]. The robots may have some agreement on the directions and orientations of local coordinate axes. They may share a common handedness or chirality (clockwise direction).

Three types of basic models are used. These models are defined according to the schedules of the operations and activation of the robots. *Asynchronous* (*ASYNC or CORDA*) model [13] is the most general one in which robots are activated arbitrarily and independently of each other. The time duration of the operations by the robots is unpredictable but finite. This implies that a robot may have done its computations on obsolete data. Due to this unpredictability, the problems become more difficult to solve in this model. The second model is the *semi-synchronous* (*SSYNC or ATOM*) [15] model. In *SSYNC* model, time is discretized into several rounds and the robots are activated in these rounds. A subset of robots becomes activated all together in a round and performs their operations instantaneously in that round. During movements, a robot is not observed by other robots in the system. The subset of robots activated in a round is not known in advance. In *fully synchronous* (*FSYNC*) model, all robots become activated in all rounds. This work assumes that scheduler activates each robot infinitely often, i.e. a scheduler is a fair [4].

Under these settings, a variety of geometric problems have been studied by the researchers. These problems include *gathering, arbitrary pattern formation, circle formation*, etc. The *circle formation* problem is defined in the following manner: a set of robots, occupying positions in the Euclidean plane, should work cooperatively to occupy distinct positions on the circumference of a circle not known a priory and this should be done within finite time. The *constrained circle formation* problem requires that while solving the circle formation problem, the maximum distance moved by any robot should be minimized.

1.1 Earlier Works

In literature, different solutions for the *circle formation* problem have been proposed under different schedulers and assumptions on the capabilities of the robots. The basic objective of these works is to propose solutions which minimize the sets of capabilities for the robots. The circle formation problem is solvable in *ASYNC* model, when robots have unlimited visibility, and the solution requires no extra assumption on the capabilities of the robots. Under limited visibility model, different solutions have been proposed with different sets of assumptions. Sugihara and Suzuki proposed a heuristic algorithm to form a circle of given radius under limited visibility [14]. Dutta et al. solved the circle formation problem for robots represented as unit discs (*fat* robots) under limited visibility model [7]. *Uniform circle formation* is another variation of the circle formation problem in which robots are asked to place themselves on the boundary of a circle such that they are equally spaced from each other. Suzuki and Yamashita proposed an algorithm for uniform circle formation for non-oblivious robots [16]. Defago and Konogaya showed that in *SSYNC* model, it is possible to converge towards a uniform circle [5]. Flocchini et al. solved the uniform circle formation problem when system has $n \neq 4$ robots [9]. Mamino and Viglietta solved the problem for $n = 4$ robots [11]. Peleg was the first to proposed the idea of using externally visible lights [12]. Das et al. characterized the computational powers of the models in which robots have externally visible lights [3]. Flocchini et al. solved the *rendezvous* problem in two different setting: (a) the robots use the lights only for remembering its own internal state and (b) they use lights to communicate with other robots its current state [10]. In memory model, solutions for the *mutual visibility* problem were also proposed [6]. A solution for the circle formation problem in the obstructed visibility model (when robots are not transparent) was proposed in [6]. None of the works in the literature have considered the constrained version of the circle formation problem.

2 Our Contribution

This work presents a study of the *constrained circle formation* problem for a set of autonomous mobile robots. The contributions of this paper are in two folds. While the circle formation problem is solvable for an arbitrary set of asynchronous robots without any extra assumption, we have shown that the constrained circle formation problem for a set of asynchronous oblivious robots is not solvable even if robots have rigid movements and both axes agreements. A characterization of the set of robot configurations for which the problem is not solvable is presented. Then, we have presented a distributed algorithm to solve the problem in admissible configurations for asynchronous robots. The algorithm uses only one bit of persistent memory. The robots do not have any form of agreements in their coordinate axis systems or chirality or constrains in movement patterns. In this weak setting, we have solved the constrained circle formation problem for asynchronous robots which use only two colours starting from an admissible initial configurations. The solution ensures collision-less movements of the robots. To the best of our knowledge, this work is the first to study the constrained circle formation problem for asynchronous robots. One of the implications of the constrained version of circle formation problem is energy efficiency.

3 General Model and Definitions

The paper considers a set of autonomous, homogeneous, anonymous, asynchronous robots under the *ASYNC (CORDA)* model. The robots are considered as points in the infinite Euclidean plane. A robot can freely move on the plane. Each robot owns a local coordinate system centred at its current position. Two distinct robots may not have same directions and the orientations of the axes and unit distances. The directions and the orientations of the axes for a robot may change with positions. Furthermore, the robots do not share a common chirality. A robot uses its local coordinate systems to locate the positions of the other robots in the system. Initially, no two robots share same point. Each robot has unlimited visibility range. A robot has non-rigid movement in which it may be stopped by an adversary before reaching its destination. However, when it moves, it moves at least a distance δ towards its destination point if it does not reach its destination where $\delta > 0$ is a constant. This assumption ensures that a robot reaches its destination within finite time. It is assumed that the robots have no knowledge about the value of δ.

- **Configuration of the robots**: Let $\mathcal{R} = \{r_1, r_2, \ldots, r_n\}$ be the set of n robots. Let $r_i(t)$ be the point occupied by r_i at time t. Let $\mathcal{R}(t) = \{r_1(t), \ldots, r_n(t)\}$ denote the robot configuration and $\widetilde{\mathcal{R}}$ be the set of all such configurations. It is assumed that in the initial configuration $\mathcal{R}(t_0)$, there is no multiplicity point (a point occupied by multiple robots).
- The closed-line segment between two points p and q includes these two points, and it is denoted by $\overline{pq}$. The open-line segment between p and q excludes these two points and is denoted by (p, q). Let $|p, q|$ denote the distance between two points p and q. Let $X \setminus Y$ denote the set difference of two sets X and Y. When measuring the angle between two line segments, we consider the angle which is has value than or equal to π.
- **Smallest enclosing circular annulus**: Let $SECA(t)$ denote the smallest enclosing circular annulus of the points in $\mathcal{R}(t)$ and $\mathcal{O}_t$ denote its centre. Let $C_{out}(t)$ and $C_{in}(t)$ denote the circles forming the outer and inner boundaries respectively of

$SECA(t)$. Let $C_{opt}(t)$ denote the circle which is equally distanced from $C_{out}(t)$ and $C_{in}(t)$ and the distance of $C_{opt}(t)$ from $C_{out}(t)$ and $C_{in}(t)$ is denoted by l_{opt}. The annular region between the circles $C_{out}(t)$ and $C_{in}(t)$ (excluding the circumferences of $C_{out}(t)$ and $C_{in}(t)$) is denoted by ANL. When there is no ambiguity, ANL, $C_{out}(t)$ and $C_{in}(t)$ are used to denote the sets of robots lying within the annular region ANL, on the circles $C_{out}(t)$ and $C_{in}(t)$), respectively. For each robot $r_i \in \mathcal{R}$, let $rad_i(t)$ denote the half line or starting from O_t and passing through $r_i(t)$.

- Let $S(t)$ denote one of the two sets $C_{out}(t)$ and $C_{in}(t)$ which contains more number of robots, i.e. $S(t) = arg\ max\{|C_{out}(t)|, |C_{in}(t)|\}$.
- **Different robot configurations**: We define the following sub-classes of $\widetilde{\mathcal{R}}$:

  - $\mathcal{E}$: A configuration $\mathcal{R}(t)$ belongs to this class if $\exists\ r_i(t), r_j(t) \in \mathcal{R}(t)$ such that either (i) $r_i(t) \in C_{out}(t)$ and $r_j(t) \in C_{in}(t)$ and $rad_i(t) = rad_j(t)$ or (ii) $r_i(t) \in C_{out}(t) \cup C_{in}(t)$ and $r_j(t) \in C_{opt}(t)$ and $rad_i(t) = rad_j(t)$.
  - $\mathcal{SR}$: It contains all configurations $\mathcal{R}(t)$ which are rotationally symmetric and $|ANL| < 3$ for $\mathcal{R}(t)$.
  - $\mathcal{CL}$: A configuration $\mathcal{R}(t)$ belongs to this class if all the robot positions in $\mathcal{R}(t)$ lie on a single line, i.e. all of them are collinear.
  - $\mathcal{M}$: It contains all configurations $\mathcal{R}(t)$ which contain at least one multiplicity point.
  - $\mathcal{H}_{\leq 7}$: A configuration $\mathcal{R}(t)$ is in this class if $|\mathcal{R}(t)| \leq 7$ and $|ANL| < 3$.
  - $\mathcal{U}$: The class is defined by $\mathcal{U} = \mathcal{E} \cup \mathcal{SR} \cup \mathcal{CL} \cup \mathcal{M} \cup \mathcal{H}_{\leq 7}$.
  - $\widetilde{\mathcal{R}}_s$: The class is defined by $\widetilde{\mathcal{R}}_s = \widetilde{\mathcal{R}}\backslash\mathcal{U}$.

To solve the constrained circle formation problem, we use the following results from pp. 163–167 of the textbook [1]:

Result 1 *For a configuration $\mathcal{R}(t)$, the smallest enclosing circular annulus $SECA(t)$ can be computed in polynomial time [1].*

Result 2 *For a nonlinear configuration $\mathcal{R}(t)$, the smallest enclosing circular annulus $SECA(t)$ has finite radius [1].*

Result 3 *For a nonlinear configuration $\mathcal{R}(t)$, the smallest enclosing circular annulus $SECA(t)$ has any one of the following properties: (i) $C_{out}(t)$ contains at least three points of $\mathcal{R}(t)$ and $C_{in}(t)$ contains at least one point of $\mathcal{R}(t)$ or (ii) $C_{out}(t)$ contains at least one point of $\mathcal{R}(t)$ and $C_{in}(t)$ contains at least three points of $\mathcal{R}(t)$ or (iii) both of $C_{out}(t)$ and $C_{in}(t)$ contains at least two points of $\mathcal{R}(t)$ [1].*

Observation 1 *The circle $C_{opt}(t)$ of a configuration $\mathcal{R}(t)$ uniquely minimizes the maximum distance from any point in $\mathcal{R}(t)$ to its circumference.*

The above observation implies that $C_{opt}(t_0)$ is the unique solution of the constrained circle formation problem for an initial configuration $\mathcal{R}(t_0)$.

Observation 2 *If a configuration $\mathcal{R}(t)$ has exactly one line of symmetry $\mathcal{L}_1$, one can define the positive direction $\mathcal{L}_1^+$ along $\mathcal{L}_1$.*

Following theorem is given without a proof:

Theorem 1 *For an initial configuration $\mathcal{R}(t_0) \in \mathcal{U}$, the constrained circle formation problem, in general, is not solvable, even if robots have persistent memory.*

4 Circle Formation Without Persistent Memory

This section presents a study of the constrained circle formation problem under a memoryless model. We provide a negative result in this setting.

Theorem 2 *The constrained circle formation problem for oblivious, asynchronous robots is deterministically unsolvable in the ASYNC model, even if robots have rigid motion.*

Proof If possible, let $\mathcal{A}$ be an algorithm which solves the constrained circle formation problem for oblivious robots. Consider an initial robot configuration $\mathcal{R}(t_0)$ as depicted in Fig. 1a. The circle $C_{opt}(t_0)$ is the desired one to be formed by the robots. All the robots should move along the line segments joining their current positions to $\mathcal{O}_{t_0}$. Now suppose that the robot r_i computes $p_i(t_0)$ and moves to this point (since scheduler is asynchronous adversary only chooses r_i for movement, the movements of the robots are rigid). This movement of r_i changes the configuration to $\mathcal{R}(t')$ as shown in Fig. 1b. Since the robots are oblivious, $\mathcal{A}$ would consider $\mathcal{R}(t')$ as a fresh initial configuration and instruct the robots to form $C_{opt}(t')$. This would cause all the robots to deviate from their original paths and would violate the optimization criteria. Hence, the theorem is true.

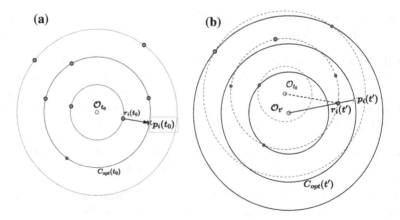

Fig. 1 An illustration of the counter-example in Theorem 2

5 Circle Formation with Persistent Memory

From observation 1, in an arbitrary initial robot configuration $\mathcal{R}(t_0) \in \widetilde{\mathcal{R}}_s$, circle $C_{opt}(t_0)$ is the unique solution for the *constrained circle formation* problem. When robots move, this circle may not remain invariant. We devise strategies in which the robots can recognize $C_{opt}(t_0)$, even if the configuration is changed. Each robot owns a single bit of persistent memory. This persistent memory is implemented via externally visible light which assumes two different colours to indicate two disjoint states. These colours do not change automatically (i.e. persistent). The lights are used with two objectives: one to store a robot's own state and other to broadcast its current state (both for communication and internal memory) [3]. A robot can identify colours of all the lights. Apart from the colour, all robots are oblivious, i.e. they do not carry any piece of information from previous cycles.

5.1 States of the Robots

Different colours of externally visible lights are used by the robots to indicate their states. Let $\mathcal{X}$ denote the set of these colours. The robots use two colours *off* and *on*, i.e. $\mathcal{X} = \{off, on\}$. The colour *on* indicates that a robot is in any one of the following states (i) active state and waiting for some other robots to turn their light *on* or to move (ii) the robot is on the circle $C_{opt}(t_0)$. The colour *off* indicates any one of the remaining states.

5.2 Algorithm Move()

Let r_i be a robot, and it wants to move to the circumference of $C_{opt}(t)$ such that the optimization criteria of the problem are also satisfied. Let $UP(t)$ be the annular region in between the circles $C_{out}(t)$ and $C_{opt}(t)$ (including the boundary of $C_{out}(t)$ and excluding the boundary of $C_{opt}(t)$) and $LOW(t)$ be the annular regions in between the circles $C_{opt}(t)$ and $C_{in}(t)$ (including the boundary of $C_{in}(t)$ and excluding the boundary of $C_{opt}(t)$). Let $C_i(t)$ denote the circle passing through $r_i(t)$ and having centre at $\mathcal{O}_t$. Let p_i be the point of intersection between $rad_i(t)$ and $C_{opt}(t)$. Let $u_i(t)$ be the intersection point between $rad_i(t)$ and $C_{out}(t)$. Let v_i be the intersection point between $rad_i(t)$ and $C_{in}(t)$. The corridor of r_i, denoted by $Cor_i(t)$, is defined as follows: (i) if $r_i(t)$ lies in $UP(t)$, then the corridor is the annular region between the circles $C_{opt}(t)$ and $C_i(t)$ (excluding the two boundaries) (Fig. 2a) and (ii) if $r_i(t)$ lies in $LOW(t)$, then the corridor is the annular region between the circles $C_{opt}(t)$ and $C_i(t)$ (including the boundary of $C_{opt}(t)$ and excluding the boundary of $C_i(t)$) (Fig. 2b). We say the $Cor_i(t)$ is free (i) for a robot in $UP(t)$ if $Cor_i(t)$ does not contain any robot position and (ii) for a robot r_i in LOW if all the robots in $Cor_i(t)$ lies on the

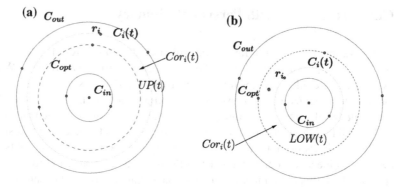

Fig. 2 An example of $cor_i(t)$: **a** r_i is in $UP(t)$ and $Cor_i(t)$ is free **b** r_i is in $LOW(t)$ and $Cor_i(t)$ is not free

circle $C_{opt}(t)$. Robot r_i moves towards the circumference of $C_{opt}(t)$ in the following way:

- **$Cor_i(t)$ is free**: If there is no robot at p_i, robot r_i moves straight towards p_i along $rad_i(t)$. Otherwise, robot r_i moves to the destination point computed in the following way:

  (i) Suppose, r_j is the robot lying at p_i. Let $\{rad_k(t), rad_s(t)\}$ be the two adjacent rays to $rad_i(t)$. Without loss of generality, suppose, $\angle r_i(t)\mathcal{O}_t r_k(t) \geq \angle r_i(t)\mathcal{O}_t r_s(t)$.

  (ii) Let d and l be two distances defined as follows: (a) if $r_i(t)$ lies in $UP(t)$, then d and l are the two distances of $r_i(t)$ from $u_i(t)$ and p_i, respectively; (b) if $r_i(t)$ lies in $LOW(t)$, then d and l are the two distances of $r_i(t)$ from $v_i(t)$ and p_i, respectively. Let $h = l + \frac{d}{2^x}$, where x is the number of robots in $(r_i(t), u_i(t))$ if $r_i(t)$ lies in $UP(t)$ or $(r_i(t), v_i(t))$ if $r_i(t)$ lies in $LOW(t)$.

  (iii) Let $\hat{C}_i(t)$ be the circle having centre at $r_i(t)$ and radius h. Let m_i be the intersection point between $C_{opt}(t)$ and $\hat{C}_i(t)$ such that m_i lies in the wedge defined by the angle $\angle r_i(t)\mathcal{O}_t r_k(t)$.

  (iv) Let $a_i(t)$ be the point on $C_{opt}(t)$ in the wedge defined by the angle $\angle r_i(t)\mathcal{O}_t r_k(t)$ such that $\angle r_i(t)\mathcal{O}_t a_i(t) = \frac{1}{3}\angle r_i(t)\mathcal{O}_t r_k(t)$. Let q_i denote the closest point among m_i and $a_i(t)$ from $p_i(t)$. The destination point of r_i is the middle point of the $arc(p_i(t), q_i)$ on the circle $C_{opt}(t)$.

- **$Cor_i(t)$ not is free**: Robot r_i does nothing.

5.3 Algorithm OptCircle()

It is assumed that (i) the initial configuration $\mathcal{R}(t_0) \in \tilde{\mathcal{R}}_s$, (ii) each robot in the system has light *off* initially and (iii) $n \geq 8$. We use result 2, result 3 and observation 1 to solve the problem. The facts stated in the result 3 are used to make $C_{opt}(t_0)$ invariant under the movements of the robots until $C_{opt}(t_0)$ becomes recognizable by the robots.

For an initial configuration $R(t_0)$, if ANL contains more than 2 robots, then all the robots in ANL compute and move to the circumference of $C_{opt}(t_0)$. Once they are on the circle $C_{opt}(t_0)$, they turn their lights *on* to make the circle recognizable to the other robots not on $C_{opt}(t_0)$. Once at least three robots on $C_{opt}(t_0)$ turn their lights *on*, the other robots compute the circle passing through the robots having light *on*, i.e. $C_{opt}(t_0)$ and move towards the circumference of the circle. Otherwise, robots are selected from the two circles $C_{int}(t_0)$ and $C_{out}(t_0)$ and are moved within ANL in such way that within finite time ANL contains at least three robots. Since the robots are asynchronous and the number of persistent lights are only two, the main challenge lies in the selection and the movements of the robots so that (i) no forbidden configuration is created due to the movements of the robots before ANL contains at least three robots; (ii) no deadlock or livelock is created during the execution of the algorithm; (iii) robots do not collide during their movements; and (iv) the annulus of the initial configuration remains same.

Let r_i be an arbitrary robot in $\mathcal{R}$. If there are at least three co-circular robots with lights *on* in $\mathcal{R}(t)$, then we are done. Robot r_i computes the circle passing through the robots having lights *on*. If r_i does not lie on this circle, it moves towards the circumference of the circle without changing its light. Otherwise, r_i does nothing. Now suppose there are less than three robots on $C_{opt}(t_0)$ having lights *on*. Depending upon the current position and configuration, robot r_i performs any one of the following actions:

- $|ANL| \geq 3$: If r_i is in ANL and $r_i \notin C_{opt}(t)$, robot r_i moves towards the circumference of $C_{opt}(t)$ according to algorithm $Move()$ and it does not change its light. If $r_i \in C_{opt}(t)$, all the robots not lying on $C_{opt}(t)$ have light *off* and $C_{opt}(t)$ contains less than three robots with lights *on*, robot r_i turns its light *on* and does not move. In the rest of the cases, r_i does nothing.
- $|ANL| < 3$: The main strategy here is to select robots from $C_{in}(t) \cup C_{out}(t)$ and move them within ANL so that ANL contains at least three robots within finite time and the annulus $SECA(t)$ remains same during the process. The robots follow algorithm $Move()$ to reach their respective destination points.

  - $r_i \in ANL$: If $r_i \in C_{opt}(t)$, robot r_i does nothing. Otherwise, it does not change its light and moves towards the circumference of $C_{opt}(t)$.
  - $r_i \notin ANL$: In this case, robot r_i lies on the boundary of the annulus $SECA(t)$. Following are the possible scenarios:
    * $|ANL| = 2$: If r_i has light *on* and there is another robot with light *on*, robot r_i moves towards $C_{opt}(t)$. Otherwise, robot r_i computes $S(t)$ and acts according to the followings:
      · $\mathcal{R}(t)$ **is asymmetric**: Since $\mathcal{R}(t)$ is asymmetric, the robot positions in $\mathcal{R}(t)$ are orderable [2]. If $r_i \in S(t)$, there is no robot with light *on* and r_i has highest order among the robots in $S(t)$, and it moves towards $C_{opt}(t)$ without changing colour of its light.
      Otherwise, robot r_i does nothing.

- $\mathcal{R}(t)$ **has one line of symmetry**: Suppose $\mathcal{L}$ is the line of symmetry. Suppose, $r_i \in S(t)$. If r_i lies on $\mathcal{L}^+$, robot r_i moves towards $C_{opt}(t)$ without changing the colour of its light. If $\mathcal{L}$ does not pass through any robot in $S(t)$ and r_i is one of the closest robots to $\mathcal{L}^+$, robot r_i turns its light *on* and does not move. In rest of the cases, r_i does nothing.

$*|ANL| = 1$: Suppose there are two robots on $C_{out}(t) \cup C_{in}(t)$ with lights *on*. If r_i has light *on*, it moves towards $C_{opt}(t)$. Otherwise, it does nothing. If there are no two robots on $C_{out}(t) \cup C_{in}(t)$ with lights *on*, robot r_i computes $S(t)$ and acts according to the followings:

- $\mathcal{R}(t)$ **is asymmetric**: If $r_i \in S(t)$ and it has highest order or the second highest order among the robots in $S(t)$, it turns its light *on* and does not move. In the rest of the cases, robot r_i does nothing.
- $\mathcal{R}(t)$ **has one line of symmetry**: If $r_i \in S(t)$, r_i does not lie on $\mathcal{L}$ and it is one of the closest to $\mathcal{L}^+$, robot r_i turns its light *on* and does not move. Otherwise, r_i does nothing.

$*|ANL| = 0$: Robot r_i computes $S(t)$. First, suppose there are two robots with lights *on*. Let $A = \{r_j, r_k\}$ be the two robots with lights *on*. Let $S_2(t) = arg\ max\{|C_{out}(t) \backslash A|, |C_{in}(t) \backslash A|\}$. If r_i has light *off* and $r_i \in S_2(t)$, it does any one of the followings: (a) $\mathcal{R}(t)$ is asymmetric and r_i has highest order among the robots in $S_2(t)$, it moves towards $C_{opt}(t)$; (b) $\mathcal{R}(t)$ has one line of symmetry and r_i is one of the closest robots to $\mathcal{L}^+$, among the robots in $S_2(t)$, it moves towards $C_{opt}(t)$. In both the cases, r_i does not change its light. Otherwise, it does nothing. Next, suppose there is at most one robot with light *on*. If r_i has highest order or second highest order among the robots in $S(t)$, it changes its light to *on* and does not move. In rest of the cases, it does nothing.

5.4   Correctness of OptCircle()

We prove that *OptCircle()* solves the constrained circle formation problem within finite time.

Lemma 1 *Algorithm Move() provides collision-free robot movements during the execution of OptCircle().*

Proof During the execution of *Move()*, robots first order themselves and then move towards $C_{opt}(t_0)$ according to that order. Thus, the robots lying on the same ray do not collide. The destination of a robot r_i lies on the $\frac{1}{3}$ section of the wedge defined by the larger angle with the neighbouring $rad_i(t)$. Thus, two robots on two different rays also do not collide. This implies that the robots have collision-free movements. Also, the destination point of a robot r_i lies within the circle having radius l_{opt} and centre at $r_i(t)$. This implies that the optimization criteria of the circle formation problem is satisfied by the movements of the robots.   □

Lemma 2 *Suppose, in a configuration $\mathcal{R}(t) \in \tilde{R}_s$, $t \geq t_0$, $|ANL| < 3$. During the execution of OptCircle(), there exists a time $t' \geq t$ such that $|ANL| \geq 3$ in the configuration $\mathcal{R}(t')$ and the circle $C_{opt}(t')$ is same as $C_{opt}(t)$.*

Proof Consider a configuration $\mathcal{R}(t)$ with $|ANL| < 3$. We prove the lemma analysing each case separately. Note that if the annulus remains same during the movements of the robots, so does $C_{opt}(t)$.

- $|ANL| = 2$: In this case, at least one and at most two robots from $S(t)$ move inside the annulus $SECA(t)$. Thus, within finite time, ANL contains at least three robots. Since $n \geq 8$ and $|ANL| = 2$, the set $S(t)$ contains at least three robots. Thus by result 3, the removal at most two robots from $S(t)$ does not change the annulus $SECA(t)$.
- $|ANL| = 1$: In this case at least two and at most three robots from $S(t) \cup S_2(t)$ move within ANL. This makes $ANL \geq 3$, in finite time. Since $n \geq 8$ and $|ANL| = 1$, the set $S(t)$ contains at least four robots. At most, two robots from the set $S(t)$ are removed. Thus, the result of 3 implies that $C_{opt}(t)$ does not change, during the movements of these robots.
- $|ANL| = 0$: In this case, at least three and at most four robots from $S(t) \cup S_2(t)$ are selected and moved inside $SECA(t)$. This makes $ANL \geq 3$, in finite time. Since $n \geq 8$ and $|ANL| = 0$, each of the sets $S(t)$ and $S_2(t)$ contains at least four robots. By the same arguments as above, the circle $C_{opt}(t)$ remains intact during the movements of these robots.

Hence, the lemma is true.   □

Lemma 3 *Suppose, in a configuration $\mathcal{R}(t) \in \tilde{R}_s$, $t \geq t_0$, $|ANL| \geq 3$. During the execution of OptCircle(), there exists a time $t' \geq t$ such that $C_{opt}(t)$ contains at least three robots with lights on, in the configuration $\mathcal{R}(t')$. Furthermore, $C_{opt}(t)$ is the unique circle in $\mathcal{R}(t')$ containing at least three robots on its circumference with lights on.*

Proof Let r_i be a robot in ANL in the configuration $\mathcal{R}(t)$. When r_i becomes active, if it finds itself on $C_{opt}(t)$, it takes any one of the following decisions (i) there is at least one robot not on $C_{opt}(t)$ with light *on* or there are at least three robots on $C_{opt}(t)$ with robot light *on*, robot r_i does nothing until the robots having lights *on* reach $C_{opt}(t)$, and (ii) all the robots, not lying on $C_{opt}(t)$, have lights *off* and $C_{opt}(t)$ contains less than three robots with lights *on*, robot r_i turns its light *on*. If r_i is not on $C_{opt}(t)$, it moves towards $C_{opt}(t)$. Thus, if $C_{opt}(t)$ contains less than three robots, within finite time, it will have at least three robots on its boundary. This implies that there exists a time $t' \geq t$ such that $C_{opt}(t)$ contains at least three robots with lights *on*, in $\mathcal{R}(t')$. The second part of the lemma follows from the case (i) and case (ii) above. Hence, the lemma is true.   □

Lemma 4 *Given an initial configuration $\mathcal{R}(t_0) \in \tilde{R}_s$, algorithm OptCircle() solves the constrained circle formation problem for a set of asynchronous robots.*

Proof By Lemma 2 and 3, there exists $t > t_0$ such that in $C_{opt}(t_0)$ is the unique circle in $\mathcal{R}(t)$ containing at least three robots on its circumference with lights *on*. Once this is done, the circle $C_{opt}(t_0)$ is uniquely recognizable by the robots even when all the robots move towards it. The robots simply compute the circle which contains at least three robots with lights *on* and then move towards it using algorithm *Move*(). Algorithm *Move*() assures collision-free robot movements. Also, during the movements, the robots satisfy the optimization criteria of the problem. Thus, within finite time all the robots reach $C_{opt}(t_0)$ satisfying the optimization criteria. Hence, the lemma is true.   □

From above results, we have the following theorem.

Theorem 3 *The constrained circle formation problem is solvable in the ASYNC model for an initial configuration $\mathcal{R}(t_0) \in \widetilde{R}_s$, when robots have externally visible lights with only 2 distinct colours.*

6 Conclusion

This paper presents a study of the constrained circle formation problem for asynchronous autonomous mobile robots. For oblivious robots, it is proved that the problem is not solvable under *ASYNC* model even when the robots have rigid movements. For robots having persistent memory, the initial robot configurations, which the problem is not solvable, are identified. For rest of the configurations, an algorithm is proposed which solves the problem for asynchronous robots which have exactly one bit of persistent memory. Following are the possible future directions of the problem: (i) relaxation of the exact optimality in the constrain considered in this work; (ii) study of the problem when robots develop faults; and (iii) the extension of the problem to the three-dimensional Euclidean space.

References

1. Berg, M., Cheong, O., Kreveld, M., Overmars, M.: Computational Geometry: Algorithms and Applications. Springer, Santa Clara, USA (2008). TELOS
2. Chaudhuri, G.S., Mukhopadhyaya, K.: Leader election and gathering for asynchronous fat robots without common chirality. J. Discrete Algorithms **33**, 171–192 (2015)
3. Das, S., Flocchini, P., Prencipe, G., Santoro, N., Yamashita, M.: The power of lights: synchronizing asynchronous robots using visible bits. In: IEEE 32nd International Conference on Distributed Computing Systems (ICDCS), pp. 506–515 (2012)
4. Défago, X., Gradinariu, M., Messika, S., Raipin-Parvédy, P.: Fault-tolerant and self-stabilizing mobile robots gathering. In: Proceeding of 20th International Symposium on Distributed Computing, pp. 46–60 (2006)
5. Défago, X., Konagaya, A.: Circle formation for oblivious anonymous mobile robots with no common sense of orientation. In: Proceedings of the Second ACM International Workshop on Principles of Mobile Computing, POMC 2002, pp. 97–104. ACM, New York, USA (2002)

6. Di Luna, G.A., Flocchini, P., Chaudhuri, S.G., Santoro, N., Viglietta, G.: Robots with lights: overcoming obstructed visibility without colliding. In: Proceeding of 16th International Symposium on Stabilization, Safety, and Security of Distributed Systems (SSS 2014), pp. 150–164 (2014)
7. Dutta, A., Chaudhuri, S.G., Datta, S., Mukhopadhyaya, K.: Circle formation by asynchronous fat robots with limited visibility. In: Proceedings of International Conference on Distributed Computing and Internet Technology (ICDCIT), pp. 83–93 (2012)
8. Flocchini, P., Prencipe, G., Santoro, N.: Distributed Computing by Oblivious Mobile Robots. Synthesis Lectures on Distributed Computing Theory. Morgan & Claypool Publishers (2012)
9. Flocchini, P., Prencipe, G., Santoro, N., Viglietta, G.: Distributed computing by mobile robots: solving the uniform circle formation problem. In: The 18th International Conference on Principles of Distributed Systems (OPODIS 2014), pp. 217–232 (2014)
10. Flocchini, P., Santoro, N., Viglietta, G., Yamashita, M.: Rendezvous of two robots with constant memory. In: Proceeding of International Colloquium on Structural Information and Communication, Complexity, pp. 189–200 (2013)
11. Mamino, M., Viglietta, G.: Square formation by asynchronous oblivious robots. In: Proceedings of the 28th Canadian Conference on Computational Geometry (CCCG), pp. 1–6 (2016)
12. Peleg, D.: Distributed coordination algorithms for mobile robot swarms: new directions and challenges. In: Distributed Computing—IWDC 2005. Lecture Notes in Computer Science, vol. 3741, pp. 1–12. Springer, Berlin (2005)
13. Prencipe, G.: Instantaneous actions versus full asynchronicity: controlling and coordinating a set of autonomous mobile robots. In: Proceeding of 7th Italian Conference on Theoretical Computer Science, pp. 154–171 (2001)
14. Sugihara, K., Suzuki, I.: Distributed motion coordination of multiple mobile robots. In: Proceedings of IEEE International Symposium on Intelligent Control, pp. 138–143 (1990)
15. Suzuki, I., Yamashita, M.: Formation and agreement problems for anonymous mobile robots. In: Proceeding of 31st Annual Conference on Communication, Control and Computing, pp. 93–102 (1993)
16. Suzuki, I., Yamashita, M.: Distributed anonymous mobile robots: formation of geometric patterns. SIAM J. Comput. **28**, 1347–1363 (1999)

Genre Fraction Detection of a Movie Using Text Mining

Sunil Saumya, Jitendra Kumar and Jyoti Prakash Singh

Abstract Movie genre plays a significant role in recommendation system as everyone has a liking for movies of specific genres. Nowadays, a Wikipedia (or wiki) page or plot for each movie is maintained on the Web. In this chapter, we propose to use the Wikipedia movie plot for genre fraction detection using text mining techniques. For our purpose, we use the bag-of-words model as topic modeling where the (frequency of) occurrence of each word is used as a feature for training a classifier. We create the corpus for 20 genres with word frequencies 1, 5, and 15 separately. Wikipedia movie plot of 640 movies is used to evaluate the proposed system. A total of 540 movie plots are used for creating corpuses, and the rest 100 are used as a test set. The system performs best on refined corpus with word frequency 15.

Keywords Bag of words · Recommendation system · Movie genre
Wikipedia movie plot · Corpus

1 Introduction

Public movies' database such as Wikipedia movie plot provides genre information to assist searching for the movie information. The tagging of movie's genres is still a manual process which involves the collection of user's suggestions. Hence, movies are often registered with inaccurate genres. Automatic genres classification of a movie based on its Wikipedia movie plot not only speeds up the classification process by providing a list of suggestions, but also effects the result potentially to be more accurate than an untrained human. A movie is a collection of genres because

S. Saumya (✉) · J. Kumar · J. P. Singh
National Institute of Technology Patna, Patna, Bihar, India
e-mail: sunils.cse15@nitp.ac.in

J. Kumar
e-mail: jitendra.itpg15@nitp.ac.in

J. P. Singh
e-mail: jps@nitp.ac.in

© Springer Nature Singapore Pte Ltd. 2018
R. Chaki et al. (eds.), *Advanced Computing and Systems for Security*,
Advances in Intelligent Systems and Computing 666,
https://doi.org/10.1007/978-981-10-8180-4_11

167

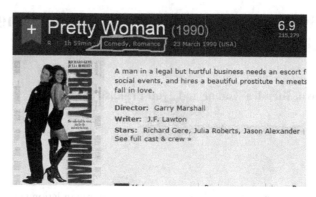

Fig. 1 A sample plot from IMDB

every genre coincides with another to form a story. People generally watch the movie of their choice. For example, some people like only action movies and some people like only romantic movies, whereas some other people like both. So, it is necessary to have right information about movie genre which helps people to like or dislike it according to their taste. Initially, when a movie releases one or more genres are associated with it which is given by movie director. Once people watch the movie, they give their own genre to that movie. Most of the time, the opinions of directors and the people match up to a certain point, but sometimes it differs. Hence, classifying correct genres of movies automatically using existing movie information is still an open research problem. This paper has established this problem and proposed a way to address it systematically. The paper not only classifies the movie genres correctly, but also detects the fraction of genres present in a movie. Suppose a movie belongs to action, comedy, and drama genres, then the fraction of genre determines how much percentage of action or comedy or drama is present. We have taken a sample plot of Pretty Woman, a Hollywood movie, from IMDB as shown in Fig. 1. The director Garry Marshall has given comedy and romance genres for this movie. Here, the fraction of comedy or romance present in the movie is not known to us. Once we calculate the fraction of each genre in a movie, it refines the user's taste. The system becomes more intelligent in identifying user's taste and gives more accurate suggestions. The implication of the research is in the movie recommendation system.

2 Literature Review

Field of cinematography has shown a considerable interest by the researcher in recent years. Experts are able to come out with substantially accurate predictions about the movie with help of readily available information about the movie. A number of sources such as the news article, blogs, or social media are present as a source for movie information.

Simoes et al. [1] explored convolutional neural networks (CNNs) in the context of movie trailers' genre classification. Results showed that proposed CNNs performed better than current state-of-the-art approaches. Pais et al. [2] addressed the automatic movie genre classification for animated movies. For each movie genre, a symbolic representation of a thematic intensity is extracted from the synopsis. A combination between the text and image descriptions was performed using a set of symbolic rules conveying human expertise. The proposed approach was tested on a set of 107 animated movies. It is observed that the text–image combination achieved precision up to 78% and a recall of 44%. On the basis of the audiovisual signal, Rasheed et al. [3] presented a method to classify movie genres. They did two layers of classifications. At the first layer, they classified each movie into action and non-action using a visual disturbance feature of each movie. Visual content measures the motion content in a clip. Next, they used color, audio, and cinematic principles for further classification into comedy, horror, drama, or others. Moncrieff et al. [4] examined localized sound energy patterns, or events, that we associate with high-level affect experienced with films. The study of sound energy events in conjunction with their intended effect enabled the analysis of film at a higher conceptual level, such as genre. Ivasic-Kos et al. [5] assumed that simple properties of a movie poster should play a significant role in the automated detection of movie genres. Therefore, low-level features based on colors and edges were extracted from poster images and used for poster classification into genres. Parkhe et al. [6] focused on genre-specific aspect-based sentiment analysis of the movie reviews. The opinion conveyed by the user toward a movie could be understood by doing sentiment analysis on the movie review text [7]. Using the aforementioned dataset and considered movie genres such as action, comedy, crime, drama, and horror, they developed a fine-grained unsupervised analysis model using lexicons that are context specific to each genre under consideration. A similar work of finding sentiments of English and Hindi text was proposed by Singh et al. [8]. Kim et al. [9] proposed the recommender system using the genre similarity and preferred genre. Using the Pearson coefficient, they found the relationship between the genres and hence created a group by k-means clustering. They evaluated the presented approach by using MovieLens dataset. Huang et al. [10] proposed a movie genre classification system using a meta-heuristic optimization algorithm called Self-Adaptive Harmony Search (SAHS) to select local features for corresponding movie genres. They extracted 277 features from each movie trailer including audio and visual features. The experimental results showed that the overall accuracy reaches 91.9%. This demonstrated more precise features can be selected for each pair of genres to get better classification results.

3 Data Collection and Methodology

3.1 Data Collection

The objective of this research is to predict the genre fraction or percentage present in a particular movie. We took Wikipedia movie plot as an input for making the dictionary of different genres. The Wikipedia movie plot is freely or easily available, and it briefly explains the movie information. As Wikipedia describes the story in brief, sometimes, words describing the story also indicate the genre information of the movie. This led us to text mining [11] approach. Once the data was collected, we did preprocessing of those data like stop word removal, pos-tagging, and stemming. At last, we create the dictionary from the preprocessed data. Figure 2 shows the flow diagram from data collection to dictionary/corpus creation.

3.2 Methodology

The proposed system initially selects a list of movies of the same genre. After the initial steps of preprocessing, the system creates the dictionary of selected movies of a particular genre. The dictionary creation is broadly classified into different modules.

Extraction and Refining Data: We extracted the movie plot from Wikipedia using the Python API for a particular movie. We used Python stop words module present in

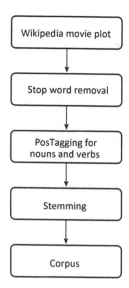

Fig. 2 Flowchart for corpus creation

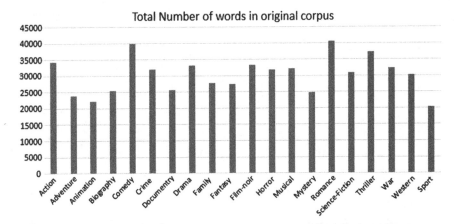

Fig. 3 Number of words in different corpus

NLTK package to remove the stop words from the extracted movie plot. A stop word is a commonly used word such as a, an, and the. As they do not contain significant information in describing the movie's genre, we removed it. We removed the proper nouns from the movie plot as proper nouns have very less significance in deciding the genre. We considered a verb for pos-tagging. We even removed adjective and adverb as they only used to improve verb quality. Next, we did stemming using PorterStemmer() method available in NLTK package. The last step was to filter those words having a word length less than 3, as these words have no significance on genre detection. We repeated this process for different genres. Dictionary name along with the number of words in it is shown in Fig. 3.

We plotted a graph to see the distribution of words and their frequencies in a corpus or dictionary. For this, first we considered *romance* corpus, which contains 40433 words. Figure 4 shows the distribution of words in a *romance* corpus. Later, we checked this distribution for all corpuses. The graph for other corpuses has not been shown here due to space limitation. As shown in Fig. 4, graph follows an exponential distribution. Next, we refined the dictionaries by fixing (i) frequency of words ≥5 and (ii) frequency of words ≥15.

Dictionary refinement and number of words in it: To refine the dictionaries, first we took those words having frequency ≥5. The number of words in each corpus has decreased after applying the word frequency ≥5 and it is shown in Fig. 5.

We repeated the same process for word frequency ≥15. The distribution of words in each corpus is shown in Fig. 6.

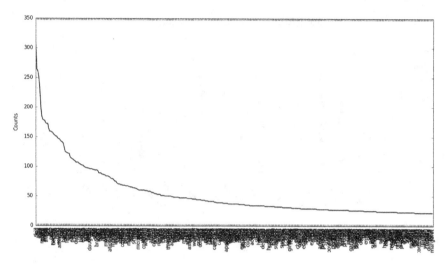

Fig. 4 Distribution of word in romance corpus

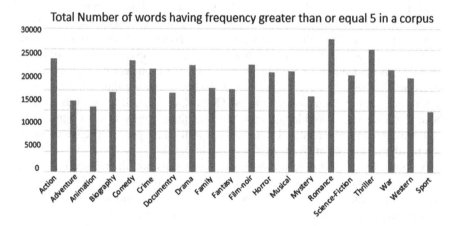

Fig. 5 Number of words in different corpus when word frequency is ≥ 5

4 Results

To find the fraction of genres, we took 640 movies plot of Hollywood, Bollywood, Korean, and Chinese movies from Wikipedia. Out of 640 movie plots, 540 were used for corpus creation. Rest 100 movie plots were used as a test set. Next, for each movie plot in the test set, we create a file which is a collection of nouns and verbs in that plot. Then, we performed the intersection between each test file and 20 corpuses created using the process described in Fig. 2, individually. Top four results were extracted and stored as final results for that test file using the process described in Fig. 7.

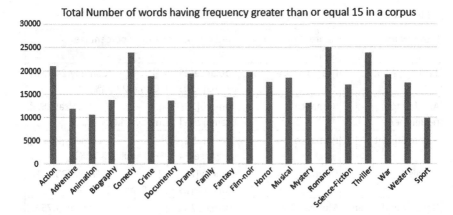

Fig. 6 Number of words in different corpus when word frequency is ≥15

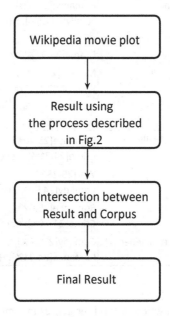

Fig. 7 Flowchart for testing

4.1 Result on Original Corpus

Table 1 shows the result after intersecting 100 test files with 20 original corpuses, where the word frequency is ≥1. Table 1 shows the genre mentioned in wiki and top four genres predicted by proposed approach. For the movie *12 Angry Men*, the genre mentioned in the Wikipedia plot is *Drama* and *Crime,* whereas predicted genres are *romance [25.25%], comedy [24.22%], drama [23.19%], and crime [27.33%]*. The

Table 1 Result on the original corpus

Movie Name	Genre mention in wiki	Predicted genre and its percentage
12 Angry Men	Drama, Crime	Romance[25.25%], Comedy[24.22%], Drama[23.19%], Crime[27.33%]
An Afair to remember	Romance	Romace[25.39%], Musical[25%], Drama[25%], Comedy[24.60%]
Before Sunset	Romance	Drama[25.2%], Comedy[25.2%], Romance[24.79%], Family[24.79%]
Boogie Night	Drama	Comedy[25.12%], Family[25.12%], Romance[25.12%], Drama[24.63%]
Dancing in the rain	Musical, Romance	Drama[27.72%], Romance[24.75%], Musical[23.76%], Family[23.76%]
Knock Up	Romance	Romance[26.02%], Musical[24.65%], Family[24.65%], Drama[24.65%]
31	Horror	Comedy[25.23%], Documentry[24.92%], Drama[24.92%], Sci-fi[24.92%]
Airlift	War,Thriller	Musical[25.17%], Romance[25.17%], War[24.82%], Drama[24.82%]
Baaghi	Action	Thriller[25%], Romance[25%], Crime[25%], War[25%]
Blood Father	Action,Thriller	Romance[25.46%], Comedy[25.09%], Crime[24.71%], Action[24.71%]
Don't Breathe	Horror	War[25.28%], Sci-fi[24.90%], Comedy[24.90%], Horror[24.90%]
Imperium	Thriller	Romance[25.02%], Adventure[25.08%], Crime[24.74%], Comedy[24.74%]
InferNo	Mystery, Thriller	Comedy[25.26%], Sci-fi[24.91%], Romance[24.91%], Horror[24.91%]

percentage indicates the fraction of top four genres in a movie. Colored rows in the table show wrong genre prediction by proposed approach. For the movie *31*, genre mentioned in the wiki is *Horror*, but top four genres predicted by our system are *Comedy, Documentary, Drama*, and *Scifi* (Science fiction). We got 4 miss out of 13 movies mentioned in Table 1. On original corpus, we got the result with the error of 30.76 percent.

Table 2 Result on the refined corpus with word frequency ≥5

Movie Name	Genre mention in wiki	Predicted genre and its percentage
12 Angry Men	Drama, Crime	Romance[25.19%], Comedy[25.19%], Drama[23.62%], Crime[25.98%]
An Afair to remember	Romance	Romace[25.86%], Musical[24.71%], Drama[24.71%], Comedy[24.71%]
Before Sunset	Romance	Drama[25%], Comedy[25%], Romance[25.58%], Thriller[24.41%]
Boogie Night	Drama	Comedy[25.78%], Musical[25%], Romance[25%], Drama[24.21%]
Dancing in the rain	Musical, Romance	Drama[25.42%], Romance[24.57%], Musical[26.27%], Comedy[23.72%]
Knock Up	Romance	Romance[25.49%], Musical[24.83%], Comedy[24.83%], Drama[24.83%]
31	Horror	Romance[25.46%], Thriller[25.46%], Horror[24.72%], Sci-fi[24.35%]
Airlift	War,Thriller	Action[25.44%], Romance[25.44%], War[24.55%], Musical[24.55%]
Baaghi	Action	Comedy[25.19%], Romance[25.19%], Thriller[24.80%], War[24.80%]
Blood Father	Action,Thriller	Thriller[25.66%], Comedy[25.22%], Romance[24.77%], Action[24.33%]
Don't Breathe	Horror	War[25.28%], Sci-fi[24.90%], Comedy[24.90%], Horror[24.90%]
Imperium	Thriller	Thriller[25.44%], Romance[25.44%], Action[24.55%], Horror[24.55%]
InferNo	Mystery, Thriller	Thriller[25.44%], Action[24.55%], Romance[25.44%], Horror[24.55%]

4.2 Result on Refined Corpus

Since the original corpus contains all words having word frequency ≥1, it is a high chance to match the most of the words from a particular movie's nouns and verbs. Hence, we refined the corpus by taking word frequency ≥5 and, later, word frequency ≥15. The results from the refined corpus are shown in Tables 2 and 3.

As shown in Table 2, the results have improved. Overall, the error is reduced by 23.07% in the refined corpus with word frequency ≥5. We got the result with the error of 7.69%. Out of 13 movies, our result could not match with the Wikipedia genre for only one movie *Baaghi*. The best result we got on refined corpus with word frequency ≥15 (Table 3). We did not encounter any error in this case as all predicted

Table 3 Result on the refined corpus with word frequency ≥ 15

Movie name	Genre mention in wiki	Predicted genre and its percentage
12 Angry Men	Drama, Crime	Drama[23.68%], Comedy[23.68%], Romance[23.68%], Crime[28.94%]
An Affair to Remember	Romance	Romance[22.28%], Musical[25.90%], Drama[25.90%], Comedy[25.90%]
Before Sunset	Romance	Romance[28.57%], Musical[19.04%], Comedy[26.66%], Drama[25.71%]
Boogie Night	Drama	Family[28.30%], Drama[24.52%], Romance[23.58%], Musical[23.58%]
Dancing in the rain	Musical, Romance	Family[28.30%], Drama[24.52%], Musical[23.58%], Romance[23.58%]
Knocked Up	Romance	Romance[25.75%], Thriller[25%], Comedy[25%], Drama[24.24%]
31	Horror	Romance[25.23%], Comedy[26.16%], Horror[24.76%], Crime[23.83%]
Airlift	War, Thriller	Romance[25.82%], Musical[25.27%], Drama[24.72%], War[24.17%]
Baaghi	Action	Romance[25.67%], Thriller[25.22%], Crime[24.77%], Action[24.32%]
Blood Father	Action, Thriller	Thriller[25.86%], War[24.71%], Comedy[24.71%], Romance[24.71%]
Don't Breathe	Horror	Thriller[27.27%], Romance[24.24%], Comedy[25.45%], Horror[23.03%]
Imperium	Thriller	War[25.69%], Comedy[25.13%], Thriller[24.17%], Drama[24.17%]
InferNo	Mystery, Thriller	Thriller[22.70%], Romance[22.70%], Horror[22.16%], Sci-fi[21.62%]

genres matched with genres mentioned in the wiki. As discussed in Sect. 3.2, we do not know the actual distribution of words in a corpus. Hence, we took the threshold value of word frequency 5 and 15 by the hit-and-trial method. The actual distribution of the words in a corpus can be calculated using topic modeling techniques such as EM algorithm and LDA. This paper has not implemented these techniques and considered this for future work. We have not compared our results with any previous work as, to the best of our knowledge, there is no such literature published in this context.

5 Conclusion

The paper proposes a system which takes movie plot as an input and automatically gives the fraction of genres present in that movie as an output. For original corpus, we got the error of 30.76%, whereas for the refined corpus with word frequency ≥5, the error was 7.69%. The system did not get any error for the word frequency ≥15.

The system lacked with a few limitations: (i) On Wikipedia, some movies have a large plot and some have a small plot, so it affects the corpus size, i.e., corpus size varies with changing movie genre; (ii) the distribution of words in the corpus is not uniform. Hence, by finding the exact distribution of words the results can be improved. Further by applying topic modeling techniques, such as EM algorithm and LDA, we can improve the result. Wisdom of crowd hypothesis [12] in the form of movie review can be used as a dataset to predict the movie genre information.

References

1. Simões, G.S., Wehrmann, J., Barros, R.C., Ruiz, D.D.: Movie genre classification with convolutional neural networks. In: 2016 International Joint Conference on Neural Networks (IJCNN), pp. 259–266. IEEE (2016)
2. Païs, G., Lambert, P., Beauchêne, D., Deloule, F., Ionescu, B.: Animated movie genre detection using symbolic fusion of text and image descriptors. In: 2012 10th International Workshop on ContentBased Multimedia Indexing (CBMI), pp. 1–6. IEEE (2012)
3. Rasheed, Z., Shah, M.: Movie genre classification by exploiting audio-visual features of previews. In: 16th International Conference on Pattern Recognition, 2002, Proceedings, vol. 2, pp. 1086–1089. IEEE (2002)
4. Moncrieff, S., Venkatesh, S., Dorai, C.: Horror film genre typing and scene labeling via audio analysis. In: 2003 International Conference on Multimedia and Expo, 2003, ICME'03, Proceedings, vol. 2, pp. II–193. IEEE (2003)
5. Ivašić-Kos, M., Pobar, M., Mikec, L.: Movie posters classification into genres based on low-level features. In: 37th International Convention on Information and Communication Technology, Electronics and Microelectronics (MIPRO) (2014)
6. Parkhe, V., Biswas, B.: Genre specific aspect based sentiment analysis of movie reviews. In: 2015 International Conference on Advances in Computing, Communications and Informatics (ICACCI), pp. 2418–2422. IEEE (2015)
7. Kumar, S., Kumar, P., Singh, M.P.: A generalized procedure of opinion mining and sentiment analysis. In: Conference on Recent Trends in Communication and Computer Networks (ComNet 2013), pp. 105–108. Elsevier (2013)
8. Singh, J.P., Rana, N.P., Alkhowaiter, W.: Sentiment analysis of products' reviews containing English and Hindi texts. In: Conference on e-Business, e-Services and e-Society, pp. 416–422. Springer (2015)
9. Kim, K.-R., Lee, J.-H., Byeon, J.-H., Moon, N.-M.: Recommender system using the movie genre similarity in mobile service. In: 2010 4th International Conference on Multimedia and Ubiquitous Engineering (MUE), pp. 1–6. IEEE (2010)
10. Huang, Y.-F., Wang, S.-H.: Movie genre classification using svm with audio and video features. In: International Conference on Active Media Technology, pp. 1–10. Springer (2012)
11. Singh, J.P., Irani, S., Rana, N.P., Dwivedi, Y.K., Saumya, S., Roy, P.K.: Predicting the "helpfulness" of online consumer reviews. J. Bus. Res. **70**, 346–355 (2017)
12. Saumya, S., Singh, J.P., Kumar, P.: Predicting stock movements using social network. In: Conference on e-Business, e-Services and e-Society, pp. 567–572. Springer (2016)

Author Index

© Springer Nature Singapore Pte Ltd. 2018
R. Chaki et al. (eds.), *Advanced Computing and Systems for Security*,
Advances in Intelligent Systems and Computing 666,
https://doi.org/10.1007/978-981-10-8180-4